The PLC Workbook

The PLC Workbook

Programmable Logic Controllers made easy

K. CLEMENTS-JEWERY

W. JEFFCOAT

PRENTICE HALL

London   New York   Toronto   Sydney   Tokyo   Singapore
Madrid   Mexico City   Munich

First published 1996 by
Prentice Hall International (UK) Limited
Campus 400, Maylands Avenue
Hemel Hempstead
Hertfordshire, HP2 7EZ
A division of
Simon & Schuster International Group

Typeset in 10/12pt Times
by Keyword Typesetting Services Ltd, Wallington, Surrey

Printed and bound in Great Britain by
T. J. Press (Padstow) Ltd

Library of Congress Cataloging-in-Publication Data

Available from the publisher

British Library Cataloguing in Publication Data

A catalogue record for this book is available from
the British Library

ISBN 0-13-489840-0

---

1 2 3 4 5 00 99 98 97 96

Contents

Preface

The PLC Workbook has been developed as a result of the experiences gained through providing training courses to industry and academe over many years and also designing, programming, installing and supporting programmable logic controllers (PLCs) in a variety of applications within a variety of industry sectors. The book will provide an invaluable guide to all engineers who are

■ involved in maintaining machines which are controlled by PLCs;
■ responsible for specifying machines and/or machine controls or
■ involved in designing, building and installing machines and/or machine controls.

The emphasis is on the practical application of PLCs in machine and equipment control and the book makes no attempt to explain the detailed construction and operation of the PLC hardware or the complexities of the software which provides its functionality. Similarly, the principles and theories governing control algorithms are not investigated.

The PLC Workbook assumes very little prior knowledge of machine control, electronics or computers and by means of simple explanations, worked examples and practical exercises it gradually leads the reader from the rudiments of control system components to a reasonable level of PLC competency.

After completing the book and carrying out all the practical exercises therein and assuming that the reader has a good engineering knowledge, he or she will be able to understand the operation of, specify, procure, design, install, operate and debug small to medium-sized PLC installations.

The reader will be provided with a basic knowledge of the subject which will enable him or her to progress towards more advanced PLC applications and programming techniques.

How to use *The PLC Workbook*

The book is structured such that it caters for a range of reader experience, from the complete newcomer to the subject of PLCs, or indeed plant and machine control systems, to the reader who has operated/installed/maintained plant or machinery that is PLC controlled. It assumes no prerequisite knowledge of the subject save a basic understanding of engineering principles, methods of production and production processes.

Chapters 1 to 4 introduce the reader to PLCs, basic elements of control and then, through simple examples and exercises, the ability to configure, design logic circuits and program the PLC is gained. In order to assist the learning process a number of exercise rigs are described. These are based upon actual rigs that have been used for training courses over a number of years and have proved invaluable in allowing students to test the theory of PLC control in a 'hands on' way.

After this gradual introduction and basic grounding in the use of PLCs, Chapter 5 begins to introduce the reader to more PLC functions and programming techniques. Chapter 6, 7 and 8 are devoted to explaining the physical parts and operation of a PLC in more detail, the kind of functions available, the devices that are used to program a PLC and how to go about selecting a PLC for a specific application. These chapters are therefore very hardware biased and provide the reader with a good understanding of PLCs and the language adopted by suppliers and users. Chapter 9 in particular gives a practical guide to any reader that may be involved in selecting a PLC for a new or replacement application. The subsequent chapters (10 and 11) deal with the support documentation that is required for PLC systems and methods of diagnosing faults and maintaining those systems.

Chapter 10 also introduces the reader to more advanced programming techniques. This chapter is intended to make the reader aware of the techniques and facilities available that can be used to increase the power and ease of operation of PLCs.

Chapter 12 consists of further exercises with sample solutions and, finally, a PLC worked example is shown in the Appendix.

As a result of working through the book, the reader will have gained a reasonable degree of understanding and competence in the design, selection, programming and maintenance of PLC systems. All that is needed to allow the reader to use this newly gained knowledge is some familiarization with the programming language that is specific to the PLC used in a particular application. This can be gained either by reading the relevant user manual or by attending the appropriate training course, a range of which are normally offered by PLC suppliers.

Acknowledgements

The authors would like to thank the following organizations for their help in providing information and photographs:

Lucas Engineering and Systems Limited.
The Scottish Maintenance Centre Limited.
Siemens Ltd.
Cutler Hammer Ltd.

CHAPTER ONE

Introduction and background

QUESTIONS ANSWERED

- What is a PLC?
- What does it consists of?
- Why use a PLC?

This chapter seeks to introduce the programmable logic controller, and provide a little background information on its main application areas and advantages.

1.1 Introducing the programmable logic controller

Programmable logic controllers, which we will refer to as PLCs, have been used in industry since 1969 and since this time they have become firmly established as the most popular means of controlling the operation of plant and machinery. They have of course evolved in terms of hardware and software. Since around 1974 micropro-cessors have been used as the very 'brain' of the PLC and this, along with the advances in electronic circuits and components, has enabled cheaper, smaller, more powerful and more reliable units to be developed. Similarly, and sometimes as a result of the hardware advances, the means of programming, the functionality available, communication features, the means of documenting programmes and fault finding have all been enhanced so that modern day PLCs are vastly superior to those encountered in the 1970s.

What, then, is a PLC? The schematic diagram (Figure 1.1) shows the basic elements of a PLC and how it is applied in its simplest form.

A PLC-controlled system consists of:

- a power supply
- an input device such as a limit switch, push button, sensor, etc.
- an input module, which is part of the PLC
- a logic unit, which is the 'brains' within the PLC

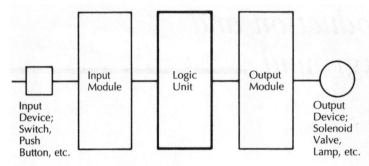

Fig. 1.1 **The basic elements of a PLC.**

- an output module, which is also part of the PLC
- an output device such as a solenoid valve, lamp, relay, motor starter, etc.

The PLC is in essence a device that is specifically designed to receive input signals and emit output signals according to the program logic. PLCs come in many shapes and sizes from small, self-contained units with very limited input/output capacity to large, modular units that can be configured to provide hundreds or even thousands of inputs/outputs. Figures 1.2 and 1.3 show a number of different PLCs.

Fig. 1.2 **Siemens 950 PLC. Reproduced with permission of Siemens.**

Fig. 1.3 **Siemens 115 PLC. Reproduced with permission of Siemens.**

1.2 Hardwired circuits versus PLC control

Prior to the introduction of programmable logic controllers most plant and machine control applications used mechanical devices such as switches and relays. Figure 1.4 shows a very simple, hardwired circuit which can be encountered around the house, in a car, in the office, etc.

This type of simple, hardwired circuit contains very straightforward 'logic', i.e. when the switch is on, the lamp is on and when the switch is off, the lamp is off. The two switches and lamps are independent of each other and because they are wired in parallel, they do not affect each other.

If we now wished to introduce some more complex logic to, say, switch off lamp 1 when lamp 2 was switched on we would have to redesign and then rewire the circuit. If we then wished to change the logic and, say, only allow lamp 2 to come on if both switches were made, but to allow lamp 1 to come on if either switch was made, then we would have to redesign and rewire the circuit again. Plant and machine control systems are much more complex than this example. They normally require events to happen in a particular sequence, different modes of operation (manual, step, automatic, for instance) and interlocks so that unusual events are catered for.

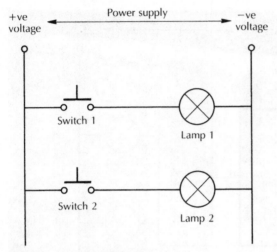

Fig. 1.4 **A hardwired system.**

Typically, a control system would consist of hundreds of relays, connected by kilometres of wire and housed in large control cabinets. Inherent in their design was the need for hundreds of connections.

Except for the use of relays as power-staging devices most of the relays in a typical control cabinet were used as logic devices, to provide a sequence of operation and control rules.

When PLCs became available it was possible to use them to take over all of the logic functions from relays and replace hundreds of relays with a more compact, solid-state unit.

Figure 1.5 shows how we could use a PLC in the simple example shown. From the diagram you can see several differences. Firstly, the switches are not connected directly to the lamps; instead the switches are connected to 'input modules' and the lamps are connected to 'output modules'. Secondly, the input modules and output modules are not connected directly to each other.

The modules are connected via the logic unit or processor only when the program conditions have been met. In the case of Figure 1.5, the logic unit is programmed to connect switch 1 to lamp 1 and/or switch 2 to lamp 2. The unit is programmed by typing a program into the processor from a keyboard. The program looks very much like an electrical ladder diagram, which will be explained in Chapter 3.

In the hardwired system the electrical current flows from the voltage source, through the switch, to the correct lamp. Electrical power simply follows the wire conductors to the lamp and when the switch is opened, the power is interrupted and the lamp goes out.

In the PLC-controlled system, electrical power comes from the voltage source, through the switch into the input module. The input module senses the presence of

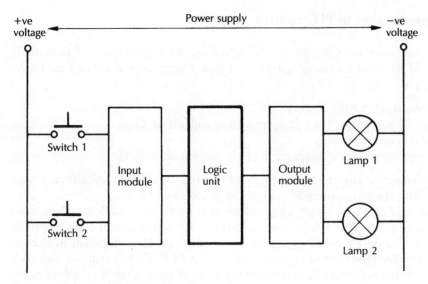

Fig. 1.5 **A PLC-controlled system.**

this voltage and, in turn, sends a small signal voltage into the logic unit or processor as it is also known. The voltage from the switch is isolated from the voltage signal that the module sends into the logic unit. This isolation is necessary since the fragile processor chip operates at very low voltage and current levels. Isolation is generally provided by an electronic component known as an 'opto coupler' which protects the processor chip.

In the example shown in Figure 1.5, the processor receives a signal from the input module when switch 1 is closed, and will send a signal to the output module as directed by the program. The program directs the signal to the appropriate output terminal for lamp 1 when it receives a signal from the input module terminal connected to switch 1. All of this takes place in typically 5–10 milliseconds.

When switch 2 is activated, a similar action is completed by the processor, but this time the signal is directed to the output module terminal for lamp 2.

With the PLC-controlled system it would be relatively easy to change the logic of the circuit, i.e. make the switches operate the lamps in a different order or sequence, without having to redesign or rewire the circuit.

Initially, due to the relatively high cost of PLCs, their application was limited to rather large control systems where the number of relays that they could replace justified their initial cost. The cost of PLCs has drastically reduced over the years and it is very cost-effective to use them for many control applications; however, it is unlikely that the example shown would warrant the use of a PLC as the circuit is so simple and unlikely to change.

1.3   Advantages of PLC control

Although the relay is inherently a reliable device, the combination of potentially hundreds of relays in a control system housed in a large control cabinet led to the following problems:

- The physical size of the cabinet
- The cost of purchasing the relays, mounting and wiring them
- The total system reliability
- The complexity of the system, particularly when fault finding.

In addition to this, any changes to the plant or machine operating sequence/logic would involve the physical repositioning and rewiring of relays.

Diagnosing faults and rectifying problems was particularly difficult when faced with the large numbers of relays, wires and connections in a typical control system.

The PLC was developed to overcome many of the problems inherent in electro-mechanical relay type control systems. As the cost of PLCs has reduced and their functionality and reliability have increased, they have taken over from relays as the most widely used means of controlling plant and machines.

A PLC can replace all the relays that would have been used to provide control logic. It is compact and easily mounted in a much smaller cabinet, requires much less wiring and because the logic is contained within its software program, changes can be implemented much more easily.

1.4   Exercise 1.1

Redraw the circuit shown in Figure 1.4 to give the following control logic:

(a)   Both switches 1 and 2 have to be pressed to operate lamps 1 and 2 at the same time. Pressing only one switch will not light either lamp and the lamps cannot be operated independently.

(b)   Either switch 1 or 2 will operate both lamps 1 and 2. The lamps cannot be operated independently and pressing both switches will not operate either lamp.

Worked sample solutions can be found in section 1.6.

1.5   Summary

Programmable logic controllers (PLCs) have been used in industry since 1969; their cost has now drastically reduced and their functionality and performance have greatly improved. The basic PLC system consists of:

- a power supply
- input and output devices

- input and output modules
- a logic unit
- a memory

and it is designed to receive input signals and emit output signals according to its program logic.

PLCs have taken over the logic functions from relays in plant and machine control applications as they are:

- more compact
- cheaper in most applications
- more reliable
- easier to fault\ find and maintain
- easier to change sequence or logic.

Overall, they provide a much better means of controlling plant and machines than electromechanical devices.

1.6   Sample solutions to exercise 1.1

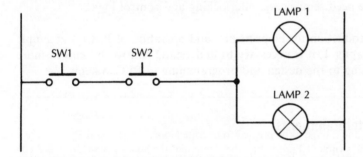

Fig. 1.6 **Solution to exercise 1.1(a).**

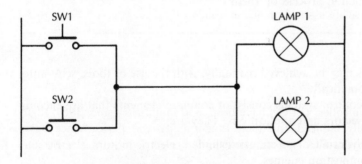

Fig. 1.7 **Solution to exercise 1.1(b).**

CHAPTER TWO

Control system basics ■■■

QUESTIONS ANSWERED

- What do machine control systems consist of?
- How are they controlled?
- What are relays and how do they work?
- How do electropneumatic circuits work?
- Why do we need sequencing, interlocking and control logic?

Having gained an introduction to the principles and operation of PLCs and simple control circuits in Chapter 1, it is necessary to understand control systems in more depth before progressing to the design and programming of PLC systems.

2.1   Control system elements

In order to produce something it is usually necessary to transform raw materials or components or products in some way. To achieve this we may wish to:

- perform some production process on them
- transport them
- store them and/or
- package them.

All of these operations may be achieved manually, with the use of tools, semi-automatically or fully automatically.

Plant and machine control systems consist of common elements that are encountered in all industrial sectors and applications. They are:

- Actuators, such as hydraulic or pneumatic cylinders, electric motors, electric solenoids or internal combustion engines.
- Switching devices, such as valves, relays, or contactors which direct power to the actuators.

- Sensors, such as switches, opto-electrical devices, thermocouples or proximity devices.
- Power supplies, either external or internal to the plant, which provide electricity, compressed air or high-pressure oil.
- Controllers, which take input signals from the sensors, give output signals to the switching devices, provide the plant sequence and control logic.

It is a combination of all of these elements that enables plant and machines to function and achieve a particular production operation. Figure 2.1 illustrates a very simple control system, consisting of an actuator (pneumatic cylinder), a switch (solenoid-operated valve), sensors (limit switch and push button), power supplies (compressed air and electricity) and a controller (could be a PLC).

The sequence of operation of the simple control system shown could be:

- The push button is depressed.
- The controller sends a signal to the valve.
- The valve switches compressed air to the rear of the cylinder piston.
- The cylinder extends (to perform some operation).
- The limit switch is depressed.
- The limit switch sends a signal to the controller.
- The controller removes the signal from the valve.
- The valve switches back and sends compressed air to the front of the cylinder piston.
- The cylinder retracts.

This demonstrates that even for a simple control sequence many control elements are necessary, all integrated into a control system, the 'brain' of which is the controller.

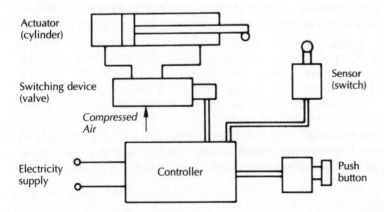

Fig. 2.1 **A simple control system.**

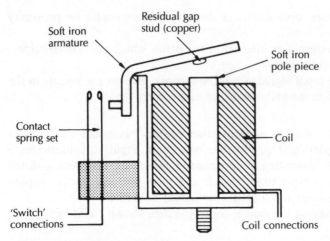

Soft iron
armature

Residual gap
stud (copper)

Soft iron
pole piece

Contact
spring set

Coil

'Switch'
connections

Coil connections

Fig. 2.2 **The basic construction of a relay.**

2.2   Electromagnetic relay control

As we have already discovered, one of the principal uses of PLCs is the replacement of electromechanical relay systems which are not programmable. It is necessary, therefore, to understand the basic operation of relays and relay control systems prior to concentrating solely on PLCs.

Relays are electrically operated switches that have a great many uses either as 'logic' devices or for use in the control of power (when they are often called 'contactors'). Their basic construction is illustrated in Figure 2.2 and consists of a coil which, when supplied with an electric current, uses the induced magnetism to attract an armature. This attraction causes a mechanical movement which operates a switch or switches.

There are a number of different types of relay available but all have the same basic principles of construction and operation. They will vary in size and shape depending upon their power-handling capability, number and type of contacts. The relay in Figure 2.2 is a normally open type with the switch being made only when the coil is energized. Other common relay types include *normally closed*, where the switch is normally closed and only opens when the coil is energized and *changeover contacts*, where power is switched from one set of contacts to another when the coil is energized.

When drawn in circuits a number of different symbols are used, a few of which are illustrated in Figure 2.3. The reader should become familiar with these symbols as they are encountered in most control system diagrams, including PLC systems.

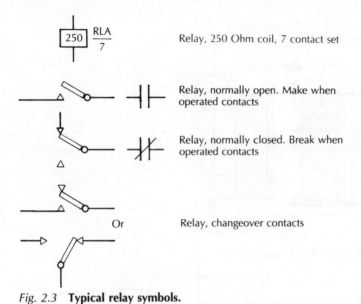

Relay, 250 Ohm coil, 7 contact set

Relay, normally open. Make when operated contacts

Relay, normally closed. Break when operated contacts

Relay, changeover contacts

Fig. 2.3 **Typical relay symbols.**

2.3   The use of relays

Armed with an understanding of the basic construction and types of relays, we can now look at some common relay applications.

2.3.1   *Power staging*

The current necessary to energize the relay coil and thus operate the armature is relatively small when compared with the current that can be handled by the relay contacts. For instance, 0.25 ampere is quite capable of switching 20 amperes through a set of contacts. A good example of such a circuit which is very familiar is the starter system of a motor vehicle, which is shown in Figure 2.4. The relay used in this application is a simple normally open, single-pole, single-throw contact. Conventionally, the relay is known as the 'solenoid' in the motor trade.

The circuit describes how very useful relays can be for switching a high current with a relatively low current. There are many more applications involving the use of relays for power staging, including three-phase supplies.

2.3.2   *Logic devices*

The concept of 'logic' as applied to control systems has already been introduced in section 1.2, where we may have wished to operate the switch and lamp arrangement

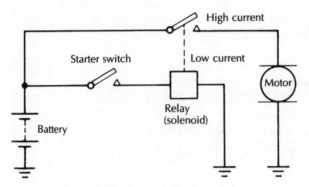

Fig. 2.4 **Motor vehicle starter circuit.**

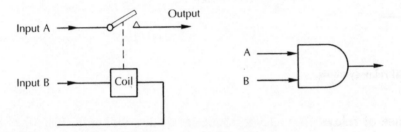

Fig. 2.5 **Part of a relay circuit and its logic symbol.**

in different ways. Control system logic is very similar to everyday, human logic as it deals with cause and effect and whether things are true or untrue.

Let us look at a very simple example. Figure 2.5 shows part of a relay circuit where we have two inputs, A and B. In order to obtain an output from the contacts both A and B have to be present (as the coil must be energized to switch the contacts). This can be shown by the logic symbol on the right-hand side. The signal fed to the A or B input could, for instance, be + 12 volts. When this signal is present we can say that a logic '1' is applied as opposed to '0' if no voltage is present.

We can now list the possible combinations of these signals for the two input 'AND' gate (this is the name given to the logic symbol as input A *and* input B have to be present for an output to be achieved).

A	B	Output
0	0	0
1	0	0
0	1	0
1	1	1

This list of combinations is known as a 'truth table'. Note that we only obtain an ouput when A and B are present together. For example, if 12 volts are applied to the

B input, the relay will energize and the contacts will close, but since no input is applied to A nothing will come out of the output. This case is listed in the third row of the truth table.

Many applications of relays as logic devices can be found in control circuits, and some are provided later in this chapter.

At this stage, it is necessary to introduce the American standard relay symbols since these are commonly used throughout the world for PLC circuit diagrams:

normally open, is shown as

normally closed, is shown as

Note that the relay is normally de-energized. The coil:

is shown as

Using these symbols, a number of relay logic circuits are described in the following sections.

2.3.3   *The latch circuit*

This circuit is a 'memory' circuit and is widely used to control (in its relay form) the power fed to machine tools. The operation of the circuit is that the operator presses the start button which starts the motor 'M' which continues to run when the start button is released, by means of the 'hold-on' circuit (Figure 2.6).

The main reason for using this circuit is that of safety. If, for instance, a fuse should blow or a power cut occur, then the 'hold-on' circuit would drop out and, in the event of the restoration of power, the operator has to restart the machine by pressing the start button again. The 'stop' button performs the same function as if the hold-on power was lost.

The method of operation of the circuit shown in Figure 2.6 is as follows:

- Press the start switch.
- Current is fed to coil A which closes two normally open contacts A1 and A2.
- One contact (A2) supplies current to the motor M.
- The other contact (A1), the hold-on contact, maintains current to the coil even though the start button is now off.
- When the power fails or the stop button is pressed (a normally on switch), the hold current ceases and the relay de-energizes, thus switching off the motor.

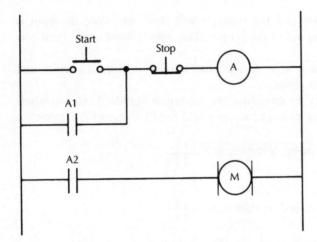

Fig. 2.6 **The latch circuit.**

2.3.4   Exercise 2.1

Figure 2.7 illustrates a relay circuit containing two relays – A with two contacts A1 and A2 and B with two contacts B1 and B2 – and two motors M1 and M2. As an exercise the reader should try to understand the sequence and write down the method of operation by applying the same reasoning as described in the latch circuit.

For the exercise assume:

- firstly, that you operate SW1 push button
- secondly, that you operate SW2, a changeover switch which is a single-pole, double-throw, spring-biased switch (shown in the rest position)
- thirdly, that you operate SW1 again and release.

How will the motors operate? The sample answer is provided in section 2.7.

2.4   Pneumatics and hydraulics

As described in section 2.1, the PLC is the controller or 'brain' of the plant or machine that is being controlled. Pneumatic, hydraulic actuators are used in profusion within automatic and semi-automatic plant and machines and so it is well worth understanding the basic operation, symbols and control problems of electropneumatic and electrohydraulic systems. An understanding of this subject will enable the reader to create suitable ladder diagrams, convert these into a PLC program and thus control such systems.

For the purposes of this book we will concentrate on pneumatic circuits as hydraulic circuits are very similar in operation.

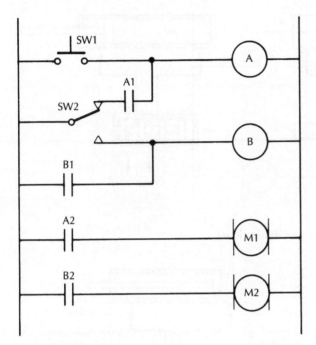

Fig. 2.7 **Circuit for exercise 2.1.**

2.4.1 *Electropneumatic valves*

The essence of electropneumatic circuits is the electropneumatic directional control valve which acts as a 'switch' to direct compressed air to each side of pneumatic actuators or to different actuators.

The five-port, two-position solenoid valve is the most commonly used valve and is shown in Figure 2.8. It is called 'five-port' because there are five drilled and tapped ports in the body of the valve, 'two-position' because the directional spool within the valve has two positions and 'solenoid' because the spool is moved back and forth by the electromagnetic actuation of solenoids (coils with soft iron cores).

To show the operation of valves in circuit diagrams, a number of 'CETOP' symbols are used. These are an internationally agreed standard for illustrating the functional operation of pneumatic components and circuits. Diagram (a) in Figure 2.8 shows the valve and cylinder combination using CETOP symbols.

The purpose of the five-port valve is to extend and retract a pneumatic piston in response to an electric signal being supplied to the appropriate solenoid, as shown in Figure 2.8. The compressed air supply is routed through the valve forcing the piston into the extended state (A+). The position of the spool in the valve is shown in diagram (b) and this was achieved by a previous operation of the left-hand solenoid.

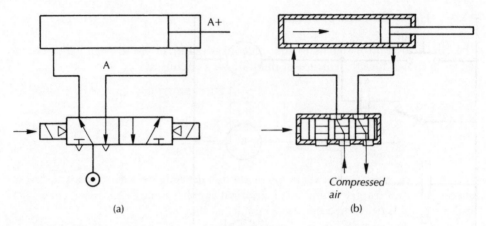

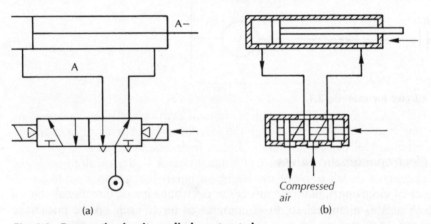

Fig. 2.8 **Basic valve and cylinder pneumatic circuit.**

(a)   (b)

Fig. 2.9 **Pneumatic circuit – cylinder retracted.**

If we now operate the right-hand solenoid, the spool shifts to the position shown in Figure 2.9. The piston is now forced to retract and remains there until the solenoid is operated again.

2.4.2   *Electrical control circuits*

The electrical control signals to the solenoid actuator of electropneumatic control valves is shown for convenience in the following circuits as:

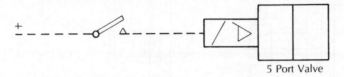

5 Port Valve

This, however, is not the conventional method of drawing electrical circuits, since the 'supply' and 'return' (+ve and −ve) electrical supplies need to be shown connected to the solenoids (to form a complete circuit) as shown:

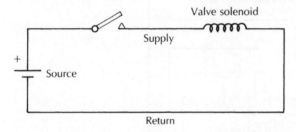

The electrical circuit for a control system is normally provided on a separate diagram from the pneumatic circuit, even though the machine control system is a combination of mechanical, electrical, pneumatic and/or hydraulic elements.

For ease of description, the electrical control circuit is drawn along with the relevant pneumatic circuit in all of the electropneumatic circuits that subsequently appear in this book.

Electrical control circuits are very similar for hydraulic applications, the basic principles being exactly the same.

2.4.3   *Multi-cylinder applications*

If we now understand the basic operation of a single cylinder controlled by a five-port valve, the next step is to extend the system to controlling two or more cylinders and to create an interlocked sequence or cycle of operation. This is the sort of application that is most likely to be encountered with plant and machine control systems, therefore we need to investigate the control problems and requirements associated with multi-cylinder operation.

Figure 2.10 shows an electropneumatic circuit with three cylinders and three valves which are five-port, two-position solenoid operated. This is the configuration used for the first exercise rig shown in Figure 2.11.

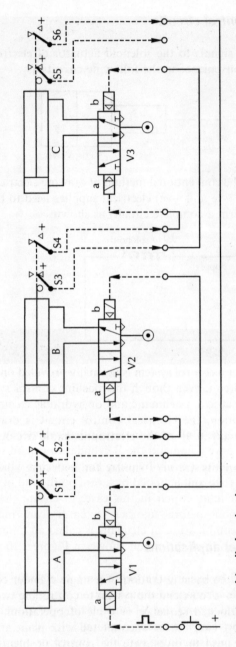

Fig. 2.10 A multi-cylinder circuit.

Fig. 2.11 **Cylinder rig.**

Referring to Figure 2.10, if we need to ensure that cylinder A is fully extended in the A + position before we command the B cylinder to extend to the B + position, this can be accomplished most easily by using limit switches which are positioned at the end of the piston stroke and actuated by a cam on the end of the piston. The use of a sensor, such as a limit switch in this case, to ensure that an actuator (the cylinder) is in position before another actuator can move is known as 'interlocking' and the combination of elements is called an 'interlock'.

The method of operation of the circuit shown in Figure 2.10 is as follows:

- The cycle is initiated by closing the switch on the left-hand side of the diagram (assume that this is a spring-release push-button type).
- This energizes solenoid 'a' on valve V1 thus pushing the spool across and allowing compressed air to the rear of cylinder A piston.
- Cylinder A extends, releasing limit switch S1 (which de-energizes solenoid 'a' on valve V2) and then depressing limit switch S2.
- As switch S2 is made, it allows an electrical current to energize solenoid 'a' on valve V2, thus pushing the spool across and allowing compressed air to the rear of cylinder B piston.

- Cylinder B extends, releasing limit switch S3 (which de-energizes solenoid 'a' on valve V3) and then depressing limit switch S4.
- As switch S4 is made, it allows an electrical current to energize solenoid 'a' on valve V3, thus pushing the spool across and allowing compressed air to the rear of cylinder C piston.
- Cylinder C extends, releasing limit switch S5 (which is not connected to any solenoid) and then depressing limit switch S6.
- As switch S6 is made, it allows an electrical current to energize solenoid 'b' on valve V1, thus pushing the spool across and allowing compressed air to the front of cylinder A piston.
- Cylinder A retracts, releasing limit switch S2 (which de-energizes solenoid 'a' on valve V2) and then depressing limit switch S1.
- As switch S1 is made, it allows an electrical current to energize solenoid 'b' on valve V2, thus pushing the spool across and allowing compressed air to the front of cylinder B piston.
- Cylinder B retracts, releasing limit switch S4 (which de-energizes solenoid 'a' on valve V3) and then depressing limit switch S3.
- As switch S3 is made, it allows an electrical current to energize solenoid 'b' on valve V3, thus pushing the spool across and allowing compressed air to the front of cylinder C piston.
- Cylinder C retracts, releasing limit switch S6 (which de-energizes solenoid 'b' on valve V1) and then depressing limit switch S5 which is not connected.

The cycle is then completed and will not start again until the push-button switch is made again. The sequence of operations can be described in simple notation as: A+ B+ C+ A− B− C−.

It would appear at first sight that to change the sequence to A+ B+ C+ C− B− A− would be just a matter of connecting up the relevant limit switches to the solenoids in the order of the desired sequence of operation.

2.4.4   Exercise 2.2

The reader should try to reconfigure the circuit diagram to give the A+ B+ C+ C− B− A− sequence. The interconnections have been left blank in Figure 2.12 to allow you to try (use a pencil to complete the diagram). If you have access to the exercise rig, then test your new circuit and see what happens before reading on. (The correct interconnections are shown in section 2.7.)

When you have attempted to carry out the exercise you will achieve the first half of the sequence (A+ B+ C+) very easily but then encounter a problem. The C− signal obtained from the operation of the C+ piston movement operating limit switch S6 is not able to achieve the movement of valve V3 spool since the C+ signal (obtained from the operation of limit switch S4 by cylinder B) is still present and must be removed before the C− movement can be achieved (since, if both solenoids are energized, the spool remains in the first position as commanded by the first solenoid

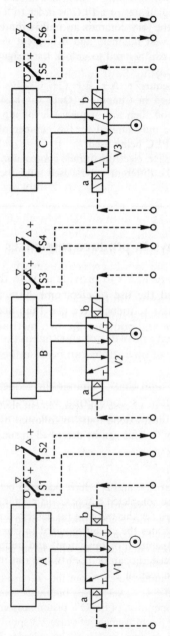

Fig. 2.12 **Exercise 2.2 – cylinder sequencing.**

to operate.) We can achieve this by rewiring the circuit but we normally need to preserve the basic wiring configuration and instead use some form of logic, using relays, electronic logic gates or a PLC in order to 'unlock' these locked up signals.

Locked signals are very common in multi-cylinder pneumatic circuits – circuits which are used in vast numbers and in a great variety of industrial automation applications. PLCs can be used to unlock these signals or carry out comprehensive interlocking functions.

This particular sequence, A+ B+ C+ C− B− A−, is achieved using a PLC program as described in Chapter 4. Once the basic concept of electropneumatic control is understood then any sequence involving any number of cylinders may be tackled – we are only limited by the number of inputs, outputs and memory available from the PLC being used.

As mentioned earlier, electrohydraulic control concepts are very similar, the main differences being the different valves used, and the need for a return line circuit configuration.

2.5   The use of relays to produce sequences

Before we look at the use of PLCs to provide us with the means of sequencing, it is quite useful to understand the use of electromechanical relays as logic devices to provide us with a particular sequence. We can then accomplish the same sequence using a PLC and compare and contrast the two methods.

2.5.1   Exercise 2.3

For the example let us choose the requirement that (referring still to Figure 2.10) cylinders A and B must extend together followed by C and then signal A and B to retract, followed by C. This can be described in simple notation as:

$$
\begin{array}{llll}
(A+) & C+ & (A-) & C- \\
(B+) &   & (B-) &
\end{array}
$$

We must ensure that A+ and B+ have occurred before C+ is permitted. Similarly, A− and B− must be completed before C− is permitted. Two relays (I and II) can be used as 'AND' gates for this purpose (as described in section 2.3.2).

Figure 2.13 illustrates the circuit required to carry out the sequence. The reader should study the operation of this circuit and understand how this interlocking is achieved. The exercise rig can be used to test out the circuit. The operation of the circuit is detailed in section 2.7.

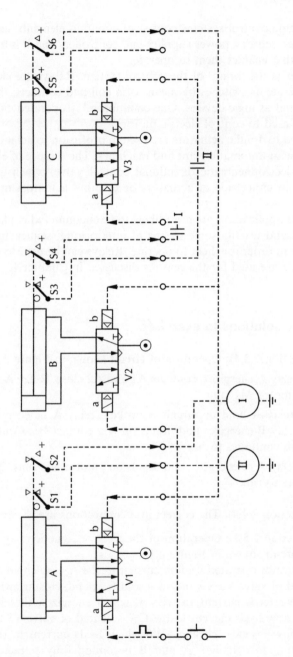

Fig. 2.13 **Example circuit using relay logic.**

2.6   Summary

Plant and machine control systems consist of common elements such as actuators, switching devices, sensors, power supplies and controllers and it is a combination of these elements that enables them to operate.

The controller is the 'brain' of the control system. Relays are electromechanical devices that operate as switches by means of a coil and contacts; they are used for power staging and as logic devices. One common relay operated circuit is the latch circuit which is used to control electric motors.

Pneumatic and hydraulic actuators are used in profusion to achieve movements in automatic and semi-automatic plant and machines. The essence of electropneumatic circuits is the electropneumatic directional control valve, which is used to direct compressed air to either side of actuators or to other actuators in response to an electrical signal.

Multi-cylinder applications are most likely to be encountered in plant and machine control systems and usually some means of interlocking, sequencing and applying logic is required in order for them to operate. Relays can be used to provide this but more often PLCs are used for the reasons discussed in Chapter 1.

2.7   Sample solutions to exercises

Exercise 2.1 (section 2.3.4): Operation of circuit shown in Figure 2.7.

Operate SW1: relay A energizes, contacts A1 and A2 close. Relay A is now 'held on' by contact A1, motor M1 runs.

Operate SW2: relay A hold on circuit is broken, relay A de-energizes, motor M1 ceases to run, relay B energizes and 'holds on' via contact B1. Contact B2 supplies motor M2 which continues to run.

Operate SW1: relay A re-energizes and 'holds on', motor M1 runs. Both motors M1 and M2 continue to run.

Exercise 2.2 (Section 2.4.4): The correct interconnections are shown in Figure 2.14.

Exercise 2.3 (section 2.5.1): Operation of the electromagnetic relay controlled electropneumatic circuit shown in Figure 2.13.

The push button is operated feeding current to the A+ solenoid of valve V1 and the B+ solenoid of valve V2. Cylinders A and B extend, closing switches S2 and S4 respectively. SW2 feeds current to relay I, which energizes, thereby closing relay contact I. SW4 now feeds current to the C+ solenoid of valve V3 through contact I. Cylinder C now extends, closing SW6 which feeds current to the A− and B− solenoids resulting in cylinders A and B becoming fully retracted. This makes

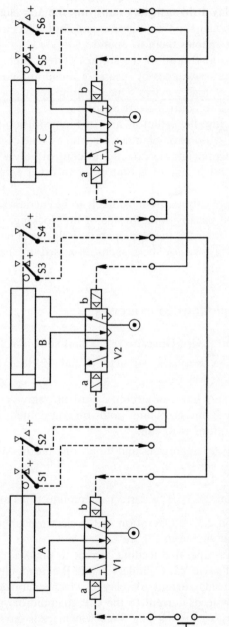

Fig. 2.14 **Solution to exercise 2.2.**

switches S1 and S3 respectively, energizing relay II. The closure of contact II will energize the C− solenoid of valve V3 giving rise to the C− motion. Note that when A− B− occurs, relay I de-energizes removing the C+ signal. This completes the cycle of operations.

CHAPTER THREE

PLC basics

QUESTIONS ANSWERED

- How does a PLC control a machine?
- What are the steps involved in using a PLC?
- Are the steps the same for any PLC?
- How do we ensure that the system is safe?
- How can we test the control system?

The previous chapters will have provided the reader with an introduction to both PLCs and control systems in general. This chapter will build upon this knowledge and show, by means of simple examples, the practical steps necessary to design a PLC system.

A good analogy for the use of PLCs in control is the function of the human brain in the control of the body. The PLC is analogous to the brain, while the pneumatic, hydraulic or electromechanical elements are analogous to the muscles.

Consider, for instance, the task of shifting a box from one location to another. Firstly, we must know where we intend to move the box and when we wish to do it. This is a 'program', i.e. a specification of what we want to do. The next consideration is for the person who is to move the box to locate its initial position. He obviously uses his eyes to do this although he could use touch (unless the room was dark then this would not be as efficient as using sight).

These messages from our senses are 'inputs' to the brain (PLC). The brain then, controlled by the 'program', instructs the arms to reach out and grasp the box and the knees to bend ready for lifting. These are 'outputs' from the brain which require the routing of power (and energy) to the appropriate muscles (actuators). The program, through the brain, now instructs other muscle groups in a coordinated sequence to move the box to the new location and release it.

This seemingly simple task that we would take for granted, involves many interacting input/output control loops, often operating in parallel, and is very complex.

From the above example the PLC can be seen as the controlling 'brain' of a control system requiring inputs, outputs and a list of instructions called 'the program' to control a particular machine.

In order to further develop the PLC control concept into a simple system, an example is given on the following page which uses the task of making tea automatically every morning for a week with one filling of a water tank per week.

3.1 The 7-day teamaker example

The reason for using this as an example is that virtually everyone knows how to make tea!

Figure 3.1 shows the system, which consists of:

- a storage tank T
- a kettle (with a heating element E)
- water valves V1 and V2
- sensors, including float switch FS and thermostat TH
- interconnecting pipework
- a time switch and an alarm bell
- and, of course, the teapot.

What, then, do we want the system to do? Make tea every morning for a week with one filling of the water tank T per week.

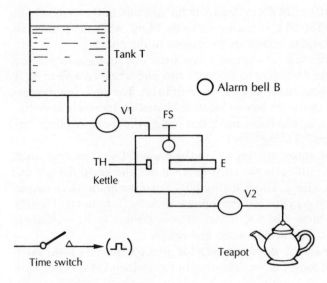

Fig. 3.1 **Diagram for the 7-day teamaker.**

3.1.1 *Operation of the teamaker*

The operation of the teamaker required in order to meet the above requirement is as follows:

- The time switch closes at the appropriate time in the morning and initiates the cycle.
- Valve V1 is opened and water fills the kettle K until the float switch FS operates.
- This switches off valve V1 and switches on heating element E.
- Water in the kettle boils and operates thermostat TH.
- This switches off the element E and switches on valve V2.
- Hot water flows into the teapot and V2 must shut off when the teapot is full.
- An alarm bell rings to inform the user that the tea is made.

The system relies upon the user to replace the teapot, complete with tea, every day and to fill up the tank each week. The sequence listed is a 'program specification'.

3.1.2 *Inputs and outputs*

A PLC is to be used to control this system and from Chapter 1 we have seen that the inputs and outputs for the PLC have to be identified prior to the design of a program.

In the teamaker example, what are the inputs?

Inputs are signals/information from sensors which inform the PLC what is happening to the system being controlled. Inputs tell the PLC what is going on. Switches, thermostats, transducers, etc., are all input devices.

Outputs are commands issued by the PLC to carry out a task (normally requiring power). Output devices have to be told when to work. Motors, solenoid valves, lamps, etc., are all output devices.

Referring to the teamaker system we can now identify each element as an input or an output device and give it a unique identification number as shown in Table 3.1.

Table 3.1 **Designating identification numbers**

Device		Letter	Identification No.
Inputs:	Time switch	TS	1
	Float switch	FS	2
	Thermostat	TH	3
Outputs:	Valve	V1	1
	Element	E	2
	Valve	V2	3
	Alarm Bell	B	4

3.1.3   Steps involved in using a PLC

From the example of the 7-day teamaker, we have so far identified:

1 The program – how the system will work.
2 The inputs and outputs – assigning a unique number to each.

We now need to carry out four more steps in order to complete the design of the PLC control system:

3 The ladder diagram design. This uses the relay logic principles that were introduced in Chapter 2 and will be explained in more detail later.
4 Programming (referred to as coding) the particular PLC with this information. This involves the use of a programming terminal and the code used is specific to the PLC being programmed. Most PLC codes are similar to each other and a short training course or study of the programming manual will normally enable the user to program the PLC (provided that the student has a good understanding of PLCs and control systems in general).
5 Testing the PLC with its program to see if it carries out the specified tasks. This is best undertaken 'off line', i.e. without the PLC being connected to any actuators that may be damaged or dangerous if the program is not working correctly.
6 Integrating the tested PLC into the plant or machine control system and testing for correct operation.

Steps 1 to 3 are general in that all PLCs need these steps and no reference has to be made to specific PLC manufacturers' instruction manuals.

Steps 4 to 6, on the other hand, are highly PLC specific as each make of PLC will have its own operating system and code for entering the program into the memory, for testing and for installation into the machine to be controlled.

3.2   Ladder diagrams

The ladder diagram is created by the application of relay logic and a degree of personal ingenuity. This can be the most difficult part of the system design, since, as in all design work, there are an infinite number of ways of achieving the desired operational specification.

3.2.1   Symbols used

The symbols used are those already discussed in Chapter 2 – the electromagnetic relay symbols. The basic symbolism is very simple, and inputs are represented as:

—| |—   i.e.   —| |—IN 1

for instance this is normally open, when a signal (i.e. a switch closing) is sent to input 1 then this closes the contact.

It is very easy to form a simple logic function; for instance, if we require to have an 'AND' gate with inputs 1 and 2 being 'ANDed' we merely put them in series:

If we require an 'OR' gate, we put them in parallel:

The 'NOT' situation (not having a signal) can be selected – this is the same as a normally-on relay contact which switches off when the relay is energized. This is shown as:

The outputs of a PLC are sometimes represented as:

(relay coils).

3.3   Circuit design

It is now an appropriate time to look at some of the control circuits that we have previously encountered and draw them in relay ladder format.

3.3.1   *Latch circuit*

The relay circuit shown in Figure 2.6, the latch or hold-on circuit, can be redrawn and inputted into the PLC as shown in Figure 3.2. Note that IN 1 is the 'start' switch, IN 2 is the 'stop' switch and OUT 1 is a relay contact which is actuated when the OUT 1 relay coil is energized.

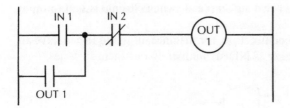

Fig. 3.2 **PLC 'hold-on' circuit.**

3.3.2   *Failsafe design of start/stop circuits*

There are two methods of producing a 'stop' signal for plant and machine control circuits:

1 Normally-open 'stop' switch (Figure 3.3).
2 Normally-closed 'stop' switch (Figure 3.4).

System 1, as shown in Figure 3.3, is unsafe because if the stop switch circuit from the switch to the PLC is an open circuit because of a fault, then no stop signal will be available and the machine will continue to operate. However, if system 2, as shown

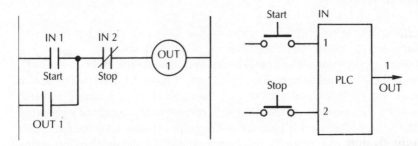

Fig. 3.3 **Normally-open circuit.**

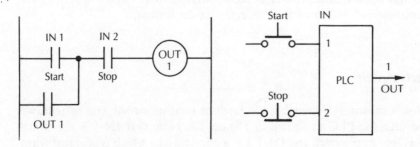

Fig. 3.4 **Normally-closed circuit.**

in Figure 3.4, is used, an open circuit will automatically cause the machine to stop – as if the actual 'stop' switch had been operated.

Safety should always be a prime consideration when designing the logic for PLC-controlled systems. A further example is now provided.

3.3.3   *Use of latch circuit to control motors*

In order to conform to health and safety guidelines for the design of systems controlled by PLCs, no safety interlocking or emergency stop signals must be dependent upon the integrity of the software within the PLC. For example, the gate switch in Figure 3.5 must be in series with the PLC output to guard against a PLC hardware or programming fault. Similarly, emergency stops must not go through the PLC program. The standard PLC arrangement for a motor control is shown in Figure 3.6.

The output latches via the starter auxiliary. If the emergency stop is pressed, or there is a motor trip or a supply fault, the auxiliary contact will be lost and the output from the PLC will de-energize (i.e. delatch). In the event of a PLC hardware fault the emergency stop will still stop the motor. (Section 6.4 covers this area in greater depth.)

3.3.4   *Timer functions*

All that is required now to enable us to design the ladder diagram for the 7-day teamaker example is an understanding of the use of timer functions.

Programs often require time delays before certain outputs come on or, conversely, for outputs to be held on for a specified length of time. Figure 3.7 describes these two timer functions. On the left-hand side of each function there is a timing diagram which shows when the output is switched on with respect to time, and the right-hand side shows the ladder diagram for each case. Diagram (a) shows an interval timer where the output is present for time 't' after the initiation of the timer by operating IN 1. Diagram (b) shows a delay 'on' operate timer where the output is only present after time 't' has elapsed, after IN 1 has been operated.

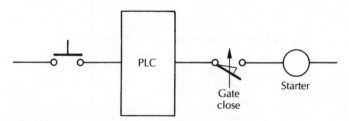

Fig. 3.5   **A PLC safety circuit.**

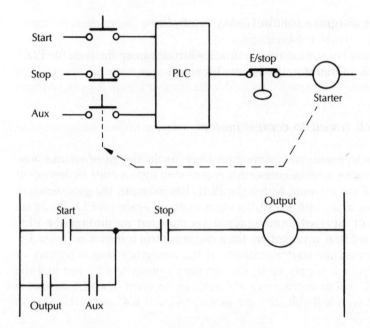

Fig. 3.6 **Standard PLC/emergency stop circuit.**

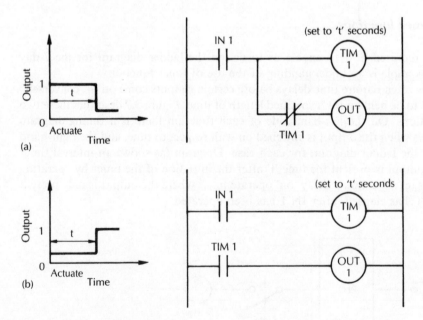

Fig. 3.7 **Timer functions.**

3.4 The ladder diagram for the 7-day teamaker

We now have knowledge of sufficient logic functions to construct the ladder diagram for the 7-day teamaker. Referring to Figure 3.1 and the sequence of operation, a ladder diagram can now be created. The ladder diagram in Figure 3.8 can be interpreted as follows:

- Output 1 (valve V1) is turned on when input 1 is triggered (the time switch) and not input 2 (the float switch).
- Output 2 (the heating element) is turned on when input 2 is triggered (the float switch) and not input 3 (the thermostat).
- Timer 1 is turned on when input 3 is triggered (the thermostat) – to allow time for teapot to fill.
- Output 3 (valve V2) is turned on when input 3 is triggered and not timer 1.
- Output 4 (the alarm bell) is turned on when the input from timer 1 is received.

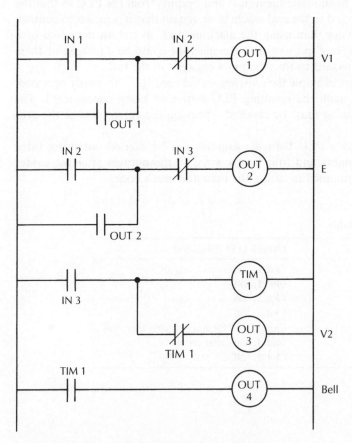

Fig. 3.8 **Ladder diagram for the 7-day teamaker.**

3.4.1   *Coding the ladder diagram into the PLC*

From the ladder diagram that has been created, the operation of entering the program into the PLC through a keyboard (or coding) is now undertaking. The details of how this is achieved and the specific coding details are unique to the make of PLC chosen for the application. Reference has to be made to the PLC manufacturer's handbook or training course.

3.5   Program test

This test is necessary after coding since a mistake may have been made during the process of keying the information into the PLC and/or by a design fault (often called a 'bug') being present but undetected by the designer.

For this test we need to simulate the inputs and outputs from the PLC so that the PLC believes it is connected to the real machine or system that it is meant to control. This is a far better practice than using the machine itself to test an unproved program, since if there is a 'bug' present then the machine could be damaged or there could even be a safety hazard to the personnel engaged in the task.

For the 7-day teamaker example the switches would need to be manually operated, as shown in Table 3.2, with the resulting PLC responses being encountered. The correct sequence and timing must be checked – proving the correctness of the program in the PLC.

If you have access to a PLC then the exercise can be carried out using three switches wired to the inputs and four lamps wired to the outputs after the ladder diagram has been programmed in, using the manufacturer's code.

Table 3.2 **Program test table**

Input switch (Operation)	Output LED (Displays)
SW1, 'on/off'	Out 1, 'on'
SW2, 'on/off'	Out 1, 'off'
	Out 2, 'on'
SW3, 'on'	Out 2, 'off'
	Out 3, 'on' for a short time then 'off'
(wait for delay)	Out 4, 'on' after delay
SW3, 'off'	Out 4, 'off'

3.5.1 *Incorrect response*

If, during the test, outputs appear when they should not or fail to appear at all, then we have to go back to the ladder diagram and analyse the reason why it does not carry out the sequence that we have specified. Sometimes, there is a need to redesign the whole diagram.

3.5.2 *Diagnostic problems*

Where much larger scale applications of PLC control are encountered, such as in industry, the complex and lengthy programs sometimes pose problems for operational and maintenance staff. This is often due to the difficulties in correlating the complex process details with the information contained in the ladder diagrams.

If, however, the examples and exercises contained in this book are worked through and understood then the basic principles of PLC applications will have been absorbed and can be applied to any system of any size.

Chapters 9, 10 and 11 contain further information and design tips.

3.6 Summary

A good analogy for the use of PLCs in control applications is the function of the human brain with the muscles being analogous to actuators and the senses being analogous to sensor inputs. The PLC requires inputs from sensors and responds with outputs according to a set of instructions called the program in order to control a machine. The steps involved in using a PLC are:

1 Determine the program specification.
2 Define the inputs and outputs and assign a unique number to each.
3 Design the ladder diagram.
4 Enter the program into the PLC memory using a programming terminal (also called coding).
5 Test the PLC program 'off line'.
6 Integrate the tested PLC into the plant or machine control system and check for correct operation.

Steps 1 to 3 are general for all PLCs but steps 4 to 6 are PLC specific.

Safety has to be a major consideration when designing and programming a PLC system and failsafe design configurations must be incorporated.

Timers are used to provide delays before certain outputs come on or, conversely, for outputs to be held on for a specified length of time.

A program test is necessary after coding since a mistake may have been made during the process of keying in the information and/or there may have been a design fault.

If, during the test, outputs appear when they should not or fail to appear when they should, then we have to go back to the ladder diagram and analyse the reason why. Where much larger scale applications of PLC control are encountered, such as in industry, the complex and lengthy programs sometimes pose problems to operational and maintenance staff. Therefore we should make every effort to design systems and programs that help the diagnosis of faults.

CHAPTER FOUR

Application of PLCs for control

- How can we use a PLC to control a sequence such as cylinders extending and retracting?
- How can we use a PLC to control a process such as mixing and heating?

In the previous chapter we have seen how a PLC could be used to control the simple example of the 7-day teamaker and the six steps necessary to design and install a PLC system. Let us now return to the much more practical example of the multicylinder circuit shown in Figure 2.10.

4.1   Cylinder sequence application

As you will recall, the task to be accomplished is to operate three pneumatic cylinders; A, B and C in a specified sequence. The example circuit shown in Figure 2.12 using two electromagnetic relays as AND gates achieved the sequence:

$$(A+) \quad C+, \qquad (A-) \quad C-$$
$$(B+) \qquad \qquad (B-)$$

Where $+$ = extend and $-$ = retract (as before).

Pistons A and B must be extended before piston C is allowed to extend; similarly for the retracting strokes.

4.1.1   The pneumatic cylinder rig

Figure 4.1 shows the circuit diagram without relays and with assigned output and input numbers, as shown in Table 4.1.

Table 4.1 **Assigned inputs and outputs**

Device		Letter	Identification No.
Inputs:	Limit switch	S1	1
	Limit switch	S2	2
	Limit switch	S3	3
	Limit switch	S4	4
	Limit switch	S5	5
	Limit switch	S6	6
	Push-button start switch	PB	8
Outputs:	LH solenoid	V1a	1
	RH solenoid	V1b	2
	LH solenoid	V2a	3
	RH solenoid	V2b	4
	LH solenoid	V3a	5
	RH solenoid	V3b	6

4.1.2   *Example sequence*

If we refer back to Chapter 3 we can ascertain that we have already carried out steps 1 (program specification) and 2 (identifying and designating inputs and outputs). What is needed now is step 3, the design of the ladder diagram. Figure 4.2 shows one possible design.

Note that we have to start the sequence by means of an additional input since as soon as the PLC is switched 'on' the cycle of events could start up on its own without a definite 'start' command.

For this diagram input 8 (IN 8) has been arbitrarily selected as the sequence start signal (we could not have chosen IN 1 to IN 6 since these have already been designated).

4.1.3   *Operation of the ladder diagram*

With reference to the ladder diagram and the electropneumatic circuit shown in Figures 4.1 and 4.2, the sequence of operation of the ladder diagram is as follows:

■ Input 8 is closed, this turns on outputs 1 and 3.
■ This operates solenoid valves V1a and V2a which causes A+ and B+ piston movements.
■ On completion of these movements switches S2 and S4 cause inputs 2 and 4 to be made.

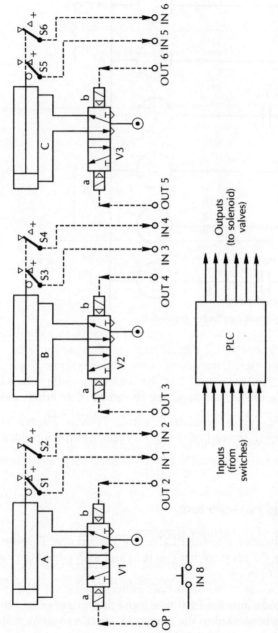

Fig. 4.1 **The cylinder exercise rig with PLC control.**

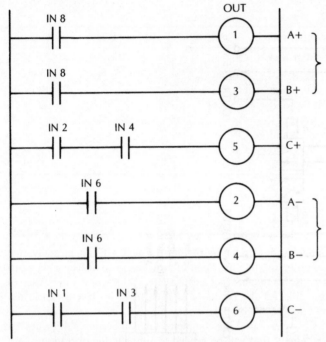

Fig. 4.2 **Ladder diagram for cylinder sequence.**

- These inputs are 'ANDed' (rung 3 of the ladder) providing an output to 5.
- This operates solenoid valve V3a giving rise to the C+ piston movement.

The remainder of the cycle is completed in similar fashion. The reader should list the remaining operations to the end of the cycle. Note that full interlocking is provided for this circuit.

4.1.4 *Coding and program test*

Coding the ladder diagram into the PLC is now carried out. This is merely a clerical operation as long as the programmer can read and understand the control functions of the particular PLC being used.

If the reader has access to the pneumatic exercise rig or any PLC he or she should input the relevant code into the PLC using an appropriate programming terminal.

Before the PLC is connected to the pneumatic rig it is to control, it is necessary to check the logic of the program and to check that no errors have been made during the coding operation. A program test schedule needs to be compiled and used for this task.

Table 4.2 **Program test table**

	Input switch		Output LEDs
Initial:	SW1 and 3	On	
Start:	SW8	On/Off	
	SW1 and 3	Off	
	SW2 and 4	On	
	SW6	On	
	SW2 and 4	Off	
	SW1 and 3	On	
	SW6	Off	

4.1.5   Exercise 4.1

As an exercise the reader should refer to Figures 4.1 and 4.2 and mentally step through the sequence, observing the operation of switches S1 to S6 being closed as the pistons move. It is possible to construct a cause–effect table which will be used to test the program. In Table 4.2 the inputs have been filled in. Use a pencil and complete the table by filling in the outputs.

Check your answers by referring to the completed table in Section 4.4. Figure 4.3 shows the 'switch box' circuit that will be needed to simulate the operation of the cylinder rig inputs and outputs.

This type of rig is useful in testing out the logic of a PLC program before it is installed in a real control system. An untested program might have some logic errors or have been typed incorrectly, which could result in electrical or physical damage to the system being controlled or, worse still, to operators or engineers working on it.

4.1.6   Exercise 4.2

Consider now the requirement for the sequence A+ B+ C+ C− B− A− referring to the circuit as shown in Figure 4.1 and as encountered in exercise 2.2. This would seem to be easy but as we found when trying to 'hardwire' the circuit it is more difficult than it seems. It would be useful to go through the process of designing the ladder diagram and observing the problems that appear as we design the ladder from scratch.

Referring to Figures 4.1 and 4.2 and starting the ladder as shown in Figure 4.4, the circuit is seen to be fully interlocked, i.e. if A+ does not occur then IN 2 is missing and thus B+ cannot happen. Carrying on with the 4th rung of the ladder:

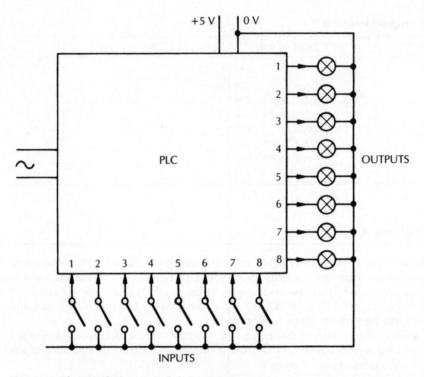

Fig. 4.3 **The simulation 'switch box' circuit.**

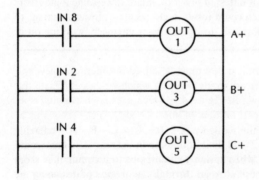

Fig. 4.4 **The first three rungs of the ladder.**

This will not work because OUT 5 is still present (IN 4 is made) and OUT 6 will not be able to override OUT 5 and cause C− to happen. (Refer to section 2.4 regarding valve actuation if in doubt.) We therefore have a 'locked' signal situation, C− cannot occur because C+ is still there. We must therefore find some method of eliminating the C+ signal once the C+ motion has been completed. A similar 'locked up' signal occurs with the B+ and B− signals.

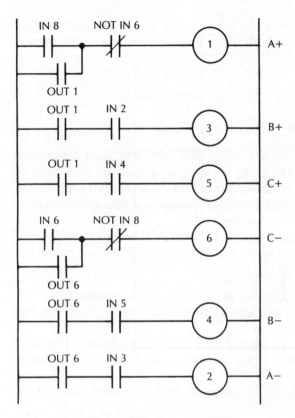

Fig. 4.5 **Ladder diagram using latch circuits.**

We have encountered a design problem that must be solved in order to achieve the desired sequence of operation. There are many ways by which we can eliminate the locked up signals. The method shown in Figure 4.5, which uses latch circuits, is as good as any and probably uses the least number of lines of code (which is economical in the use of PLC memory capacity).

Note that the outputs (OUT 1, etc.) are used in this example as if they were inputs. This is a very useful feature of PLC programming enabling us to determine the state of outputs. The reader should attempt to work through the logic of this ladder to prove the reasoning behind the design and write down how it works. A description of how the circuit works is given in section 4.4.

4.2   Control of processes by PLC

Many of examples given so far have featured the use of a PLC to control the cylinders in a particular sequence. There are very many applications where PLCs are used to control processes across the whole range of industry sectors. The following example will illustrate that a PLC can be used to control simple process plant.

4.2.1   *The process control rig*

The process control exercise rig consists of a number of tanks, level detectors, a water heater and a thermostat. The tanks are filled to a specified level prior to producing a mixture of the fluids contained in the tanks at a specified temperature. This is then pumped to a storage vessel. Figure 4.6 provides a diagram of the exercise rig and Table 4.3 shows the inputs and outputs.

(*Note*: At the start of the cycle of operation all tanks A, B and C are filled to the correct level. Referring to Chapter 3 where six steps are set out in order to enable a PLC system to be 'up and running':

- Step 1 is to establish a detailed sequence of operation that is specified.
- Step 2 is to identify the inputs and outputs relative to the PLC. Five inputs and eight outputs in this case.
- Step 3 is to design the program, a ladder diagram. The diagram given in Figure 4.7 is one of the many design solutions that may be created to achieve this desired sequence of operation.
- Step 4 is to code the ladder diagram into the specific PLC that is to be used. This is machine specific.
- Step 5 is to test the validity of the design and the accuracy of the coding operation. The switch box shown in Figure 4.3 is used for this task, the outputs created by the program being displayed by output LED displays as shown in Figure 4.8. Also shown in the figure is a timing diagram, known as a phase diagram which provides a pictorial representation of events that are described in words in section 4.2.2. Motion and phase control diagrams are more fully described in Chapter 7.

Table 4.3 **Input/output designation**

Inputs		No.	Outputs	No.
Level detector	LD1	1(⌐)	Valve V2	1
	LD2	2(⌐)	Motor M1	2
	LD3	3(⌐)	Motor M3	3
	LD4	4(⌐)	Valve V1	4
	Start signal	8(⌐⌐)	Motor M2	5
			PID heater	6
			Valve V3	7
			Valve V4	8

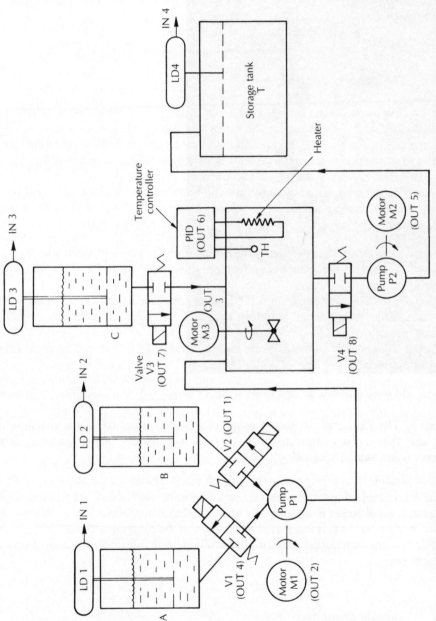

Fig. 4.6 (a) The process exercise rig.

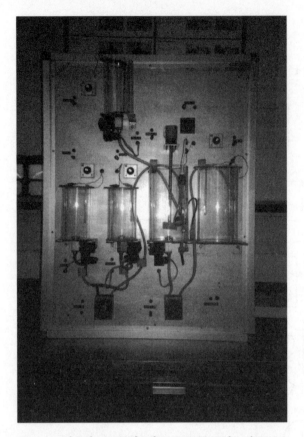

Fig. 4.6 **(b) Photograph of process exercise rig.**

- Step 6. The PLC with its tested program is now connected to the machine or process that is being controlled. A basic knowledge of electrical engineering principles is needed for this task.

It is not possible to overemphasize the importance of planned, careful testing of PLC programs. Indeed, all programs should be tested immediately they are written, while concepts are still fresh. Even after the most careful testing, however, problems can still arise when the PLC is integrated to the plant to be controlled. Such faults will be due to problems with the machine or plant itself, or with our understanding of how it will operate.

4.2.2   *Example sequence*

The task is to design and implement a PLC program to achieve the specified sequence of operation, transferring a certain quantity of fluid in tank A to the mixing

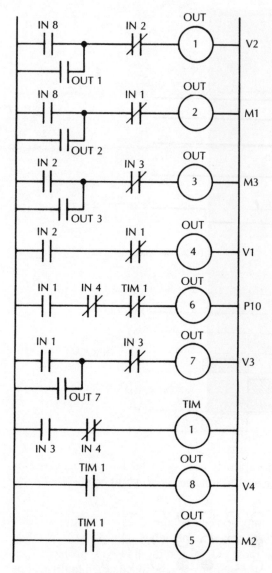

Fig. 4.7 **Ladder diagram for process exercise rig.**

tank and another measured quantity of fluid from tank B to the mixing vessel. During this transfer, the mixing motor is started in order to mix fluids A and B together.

Heat is now supplied by a small immersion heater, under the control of a PID temperature controller. PID stands for 'proportional integral differential' and is a method of achieving precise control of a process parameter such as temperature. (Further details are provided in Chapter 6.)

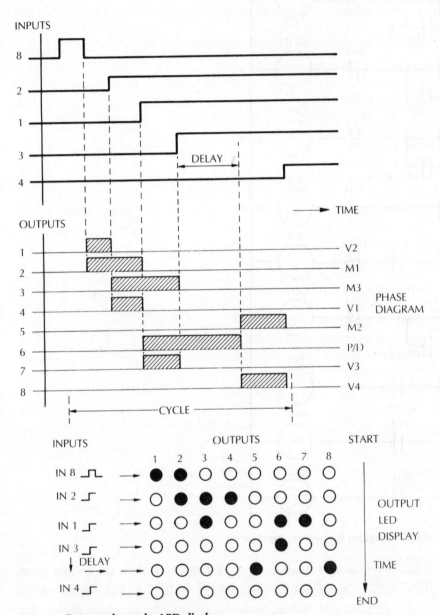

Fig. 4.8 **Outputs shown by LED displays.**

Liquid from tank C is then fed by gravity to the mixing vessel to be warmed and mixed with the fluids from tanks A and B. After a delay the composite mixture is transferred to the storage tank T. The batch cycle is now complete. The layout of this rig is shown in Figure 4.6.

The sequence of operation for the rig is as follows:

> Start (switch IN 8 is pressed)
> Open valve V2
> Start motor M1
> Wait until level detector LD2 detects 'low'
> Close valve V2
> Start motor M3
> Open valve V1
> Wait until level detector LD1 detects 'low'
> Turn motor M1 off
> Turn valve V1 off
> Turn heater PID on
> Turn motor M3 off
> Turn valve V3 off
> Delay by 10 seconds
> Start motor M2
> Open valve V4
> Turn PID heater off
> Wait until level detector LD4 detects 'high'
> Turn motor M2 off
> Turn valve V4 off
> End of cycle

Program test

Input switches (X)		Output (LEDs) (Y)
8 on/off	$\longrightarrow$	1 on, 2 on
2 on	$\longrightarrow$	1 off, 3 on, 4 on
1 on	$\longrightarrow$	2 off, 4 off, 6 on, 7 on
3 on	$\longrightarrow$	3 off, 5 on, 7 off, 8 on
4 on	$\longrightarrow$	5 off, 8 off

4.3   Summary

PLCs are most commonly used for industrial control applications such as the control of pneumatic, hydraulic and electric actuators. As the examples have shown, they can be applied to cylinder sequence control and the control of processes.

The main steps in designing and implementing a PLC solution are:

- Step 1.   List the sequence of operation.

- Step 2.   Identify and list the inputs and outputs.
- Step 3.   Design the ladder diagram.
- Step 4.   Load the program into the PLC (code).
- Step 5.   Test the program using a 'test box'.
- Step 6.   Integrate the PLC into the control system and prove out.

4.4   Sample solutions to exercises

Exercise 4.1 (Section 4.1.5)

Sequence:

A+   A−
          C+   C−
B+   B−

Program test table

	Input switch			Output LEDs	
Initial:	SW1 and 3	On		6	On
Start:	SW8	On/Off	→	1 and 3	On/Off
	SW1 and 3	Off	→	6	Off
	SW2 and 4	On	→	5	On
	SW6	On	→	2 and 4	On
	SW2 and 4	Off	→	5	Off
	SW1 and 3	On	→	6	On
	SW6	Off	→	2 and 4	Off

Exercise 4.2 (section 4.1.6): Operation of the ladder diagram in Figure 4.5

1st rung: IN 8 is made (⎍), OUT 1 'energizes', OUT 1 contact makes and latches through normally made IN 6 contact. (Note that SW6 is open because cylinder C is in the C− position). OUT 1 supplies A+ solenoid and A+ motion occurs making SW2 (= IN2).

2nd rung: OUT 1 already made (latched), IN 2 makes which 'energizes' OUT 3, which results in the B+ motion.

3rd rung: OUT 1 already made (latched), IN 4 contact made because of the B+ action. This 'energizes' OUT 5 coil and therefore results in the C+ action.

4th rung: IN 6 now makes which through the NOT IN 8 contact 'energizes' OUT 6 coil giving us the C– motion. OUT 6 contact latches the OUT 6 coil. Simultaneously NOT IN 6 contact is opened by the appearance of the IN 6 signal – which 'unlatches' OUT 1 coil and removes OUT 1 contacts from rungs 1, 2 and 3 removing in turn the power from the A+, B+ and C+ solenoids.

5th rung: When C– motion is completed, IN 5 makes which gives rise to OUT 4 (B+ action).

6th rung: When B– motion is completed, IN 3 makes giving OUT 2 which results in the A– motion.

This is the end of the cycle which can be made to continuously repeat itself if the end of the cycle (SW1 being operated) is used as the START signal for cycle to commence, i.e. as IN 8. One has only to join SW1 lead to the IN 8 PLC input for this to happen!

CHAPTER FIVE

More PLC functions and programming

QUESTIONS ANSWERED

- How can we introduce time delays into the sequence?
- How can the PLC be programmed to achieve this?
- How can we program the PLC to count for instance the number of operations of a machine?
- Why would we wish to convert a ladder diagram to logic or vice-versa?
- Why would we wish to use control relays?
- How can we achieve this?

Through the examples and exercises used in Chapter 4, we have begun to learn about the practical application of PLCs for industrial sequence and process control. We can now improve the operation and flexibility of PLCs by using some other functions and programming methods that are available to us.

5.1 Timers and counters

The timing circuits introduced in Chapter 3 can be used as a method of control instead of the use of logic elements such as latch circuits. As an example, consider the sequence A+ B+ C+ C− B− A− which was previously achieved by means of two latch circuits, as shown in Figure 4.5. We can achieve the same sequence by means of timers which only activate the solenoid for a short duration and hence prevent the locked signals from occurring. The reader will recall that it was these locked signals that originally made the latch circuits necessary.

5.1.1 *The use of timers*

The ladder diagram using only timers is shown in Figure 5.1. In order to reduce the effect of timing circuits on the overall cycle time for the control system (which will usually increase with the use of timers) it is necessary to reduce the pulse duration to the shortest possible time, for example, 0.1 to 0.5 second.

By combining timing circuits we can derive any output pulse or function we desire. For example, if we require a 3 second pulse to appear 5 seconds after initiation we could use the circuit shown in Figure 5.2. Referring to this circuit we can see that operation IN 1 sets off timers TIM 1 and TIM 2. The waveforms as shown indicate that outputs OUT 1 and OUT 2 are only present together for 3 seconds, which by 'adding' OUT 1 and OUT 2 results in the 3 second delayed pulse at output OUT 3.

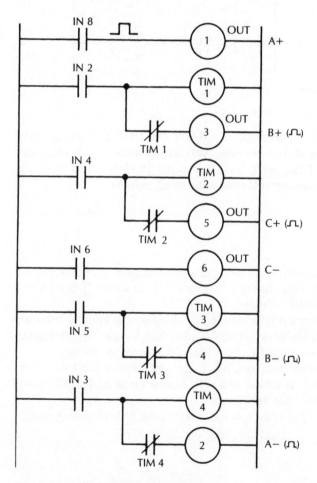

Fig. 5.1 **Ladder diagram using timers.**

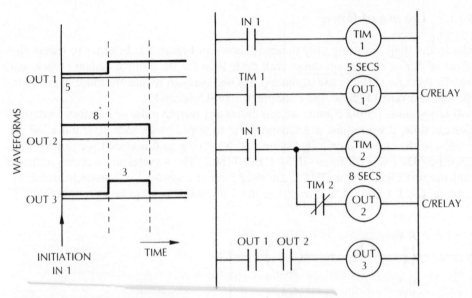

Fig. 5.2 **Combining timing circuits.**

This method of using a number of outputs, i.e. OUT 1 and OUT 2, to produce the real output OUT 3 is very wasteful of outputs. Auxiliary outputs, which are sometimes called control relays or flags, are available in many PLCs to allow us to use circuits of this type without using up real (and expensive) outputs.

5.2   Counters

Many applications in machine control call for a prescribed number of operations; for example, a cylinder strokes 12 times during a machine cycle to locate 12 bottles in a case prior to the case being shrink wrapped.

This counting function is very easily achieved by calling up the appropriate rung of the ladder diagram which includes a counter. So that for the 'bottle loader' example the ladder diagram shown in Figure 5.3 would operate as follows:

In IN 1 operates 12 times then counter CNT 1 operates, closing CNT 1 contact and giving an output at OUT 1. If a reset pulse 'R' is now input by IN 2, then the counter is reset to zero ready for the next count.

Figure 5.4 illustrates a counting circuit which achieves the following sequence:

A+ to A− repeats 15 times then B+ C+ B− C−.

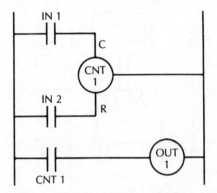

Fig. 5.3 **Ladder diagram incorporating counters.**

The cycle of operation is as follows:

   Input IN 8 (start signal) is made
   IN 1 is currently on because of the A− position
   This produces an output OUT 1
   This produces A+
   This makes IN 2 and releases IN 1, which signals the return of piston A−
      because OUT 2 is present
   This again makes IN 1 and the cycle repeats
   IN 1 also feeds pulses to counter CNT 1 which has been set to 15 (via
      the code)
   On receipt of the 15th pulse, the counter operates
   This gives output OUT 3 which makes B+
   IN 4 makes C+
   IN 4 also provides a reset pulse to the counter, and this breaks the CNT 1
      contact thus turning OUT 3 off
   The C+ action makes IN 6 which gives OUT 4
   This causes B− movement and IN 4 is lost
   The counter then starts counting again to 15.

It is important to note that when fault finding in PLC-controlled systems it is necessary to understand how the circuit works prior to attempting to locate the fault. It is very difficult to locate and fix a fault unless the method of operation is known.

5.3   Ladder/logic conversion

Ladder rungs are by no means the only method we can use to define how a control system works. Indeed some PLCs are programmed by defining the relationship between the inputs and outputs using the logical operators, AND, OR, NOT, etc.

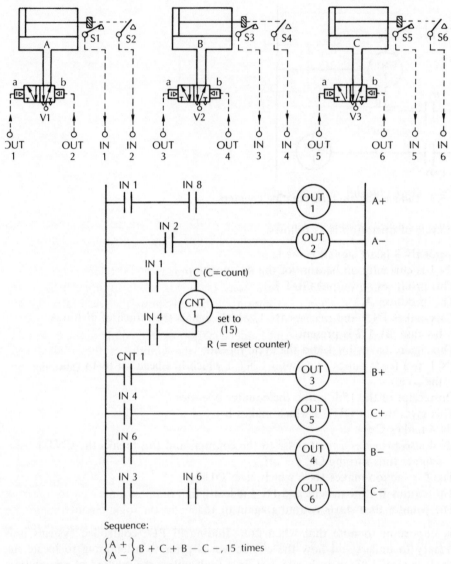

Fig. 5.4 **Cylinder circuit using counters.**

This section is intended to show how we may convert from ladder to logical symbols and vice versa.

The same basic skills of translating logic can be required to allow us to re-engineer existing control systems. Converting a system that has been designed using logic symbols (e.g. hardwired electronic control) into a PLC program, is a typical example. The converse is also true where very high speed is required, and we need to transfer part of a PLC program into dedicated high-speed electronic logic.

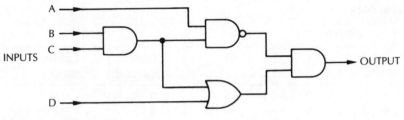

Fig. 5.5 **Logic circuit.**

5.3.1 Logic/ladder conversion

At first sight it may appear easy to replace AND and OR gates with ladder rungs but, as illustrated in the very simple example below, the conversion is not always so straightforward.

Consider the logic circuit in Figure 5.5, where four inputs produce one output.

In order to check the validity of the conversion it is helpful to use a 'truth table' (also called a Boolean algebra evaluation). Table 5.1 shows the truth table for this diagram.

If we now draw up the equivalent ladder diagram we can compare the truth table combinations of the new ladder diagram with the outputs of the original logic system. If the conversion is perfect, then of course the two output combinations should be identical.

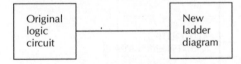

Compare using table 5.1

If this is attempted as an exercise, the reader will find that it is quite easy to construct an incorrect ladder diagram that is not equivalent to the logic circuit. Figure 5.6 provides a solution to the problem. Note the use of auxiliary outputs (control relays) which allow us to create the ladder diagram in its simplest form.

5.4 Control relays – or flags

5.4.1 Multiple inputs

A limit exists on the number of input symbols that can be in one rung of a ladder diagram, i.e. 8 inputs; thus if 10 inputs are necessary in order to produce output 1 then a control relay R1 can be used to permit this to occur. Figure 5.7 illustrates this situation.

Table 5.1 **Truth table**

| INPUTS | | | | OUT 1 | |
A	B	C	D	Logic (original)	Ladder (new)
0	0	0	0	0	
1	0	0	0	0	
0	1	0	0	0	
1	1	0	0	0	
0	0	1	0	0	
1	0	1	0	0	
0	1	1	0	1	
1	1	1	0	0	
0	0	0	1	1	
1	0	0	1	1	
0	1	0	1	1	
1	1	0	1	1	
0	0	1	1	1	
1	0	1	1	1	
0	1	1	1	1	
1	1	1	1	0	

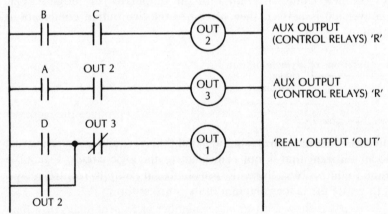

Fig. 5.6 **Conversion to ladder diagram.**

5.4.2 *Multiple operations of a single output*

If an output, e.g. OUT 1, need to be operated more than once during a cycle of operation then it is not normally permissible to have this OUT 1 shown on the ladder diagram more than once. Most PLCs will not accept this program instruction should one attempt to carry out this task, e.g. OUT 1 may need to operate twice during the

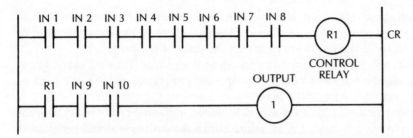

Fig. 5.7 **Extending the number of contacts via rung.**

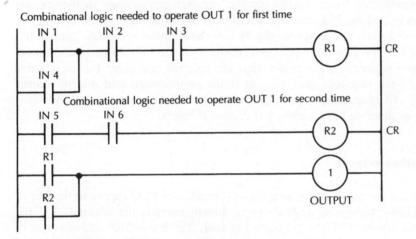

Fig. 5.8 **Multiple operations of a single output.**

cycle and we would thus use the control relays R1 and R2 to produce the required inputs suitably OR as shown in Figure 5.8.

There are many more uses of control relays that help the programmer to produce a simpler final operational program. A typical example is shown in the logic → ladder example (Figures 5.5 and 5.6) which illustrates how useful these control relays can be.

Many PLC systems often show a much larger number of control relays than real outputs – for instance, the Cutler & Hammer 40I0 model has 16 Y outputs ('real' outputs) and 128 R outputs (control relays).

5.5 Controlled reset

During an operational sequence of interlocked events it is possible that, due to the failure of a system component, the program and the physical state of the hardware become out of phase. Once the defective item is replaced we have the problem of either 'forcing' the program to advance to a point where it 'matches' the state of hardware, or to force the hardware to match the state of the program.

Both methods are difficult to execute and can be dangerous. The staff carrying out the procedure require full knowledge of the operating sequence and the safety implications of moving each part of the plant.

A safer, more efficient, and faster way is to design a manual reset facility into the control system to allow a single key switch to return the entire plant and program to a known 'home' position. Two options are open to us: the controls could be hard-wired, or designed into the PLC program. A hardwired option would work even with the PLC disconnected but it would involve complex connections to incorporate safety interlocks and the PLC outputs would need to be disabled. In the vast majority of cases extra code designed into the PLC is the simplest and safest approach.

As with all PLC programs the reset code needs to be carefully designed and tested. Particular care is necessary to ensure that the solution will cater for all possible configurations the sequence may stop at (both programmed and due to human intervention). To illustrate the principles the very simple sequence A+ B+ C+ C− B− A− as described in Chapter 4 is discussed below.

5.5.1   Hardware reset

The reset switch (normally a key switch) will 'break' the PLC supply to disable its outputs and then, through separate contacts, directly energize the solenoids A−, B− and C−. This will reset all the cylinders together. The hardwired modifications to achieve this are shown in Figure 5.9. Where a sequential (C− B− A−) return sequence is required, extra interlock switches and relays would be required, probably making a hardwired solution impractical.

5.5.2   Software reset

Figure 5.10 illustrates (as shaded lines) a simple change to the original program (Figure 4.5) which will move all the cylinders back to the original position A− B− C−. Note how the extend output rungs are 'blocked' by a normally-closed reset contact and the return rungs a 'forced' by a normally-open contact. In this implementation all the cylinders will move 'home' together when the reset switch is operated.

5.5.3   Specified return sequence programs

The 'return home' system shown in Figure 5.10 does not speciefy the order in which the cylinders move back. In many instances the reset sequence is critical to avoid

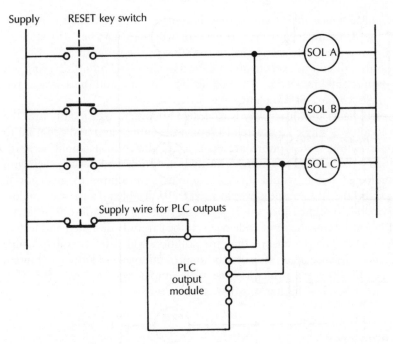

Fig. 5.9 **Hardwired reset circuit.**

physical damage to the plant or, more important, to maintain operator safety. The more complex the program the more difficult, but important, it is to define the reset sequence. If you, the system designer, cannot define a safe reset sequence, how can a maintenance engineer or operator working under pressure to restore production be expected to 'force' the plant safe?

A good starting point to define a reset sequence is simply to use the logic built into the 'auto' rungs. Figure 5.11 shows how the home limit for cylinder C is used to enable the return of cylinder B through the reset switch. In turn, cylinder B returning will enable A. It may be argued such an approach is limiting, requiring the switches to be working before the reset sequence can operate. This is true but in most cases desirable, as we will not be able to progress through the subsequent sequences until the switch is repaired.

Some programs may use counters or latched flags to keep track of the sequence progress. The reset code for such a program would also need to reset these counters and flags.

5.6 Manual controls

There are circumstances in which it is necessary to move parts of a machine out of their normal sequence to set tooling or switches. This can be achieved using the

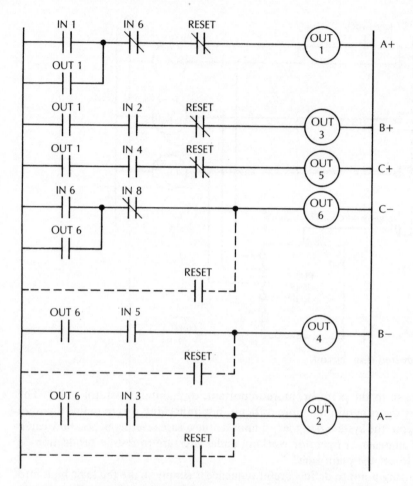

Fig. 5.10 **Simple reset code additions.**

programmer to 'force' the outputs or by operating solenoids on the plant. However, operating outputs by either of these methods overrides all the safety interlocks and can lead to serious damage to the plant or even put personnel in danger.

A better solution is to build manual controls into the PLC program for as many movements as possible and particularly for any that are dangerous or could cause damage. This code can be implemented in parallel with the auto code, using the same interlocks to prevent 'illegal' or dangerous movements. An AUTO/MANUAL mode switch is used to select the operating mode and a push button provided for each movement required. An example of manual controls added to the movements in our example from Chapter 4, Figure 4.5, are shown in shaded lines in Figure 5.12. In this example the manual interlocks have been designed to allow cylinder C to move

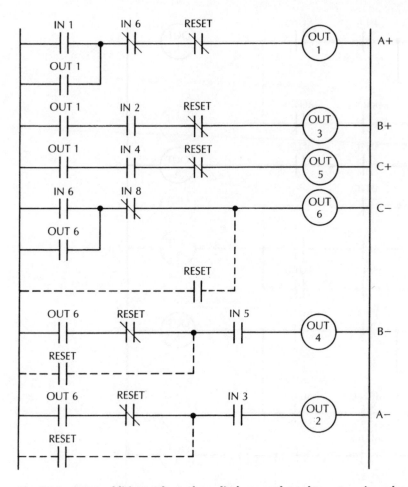

Fig. 5.11 **Reset additions where the cylinders are forced to return in order C– B– A–.**

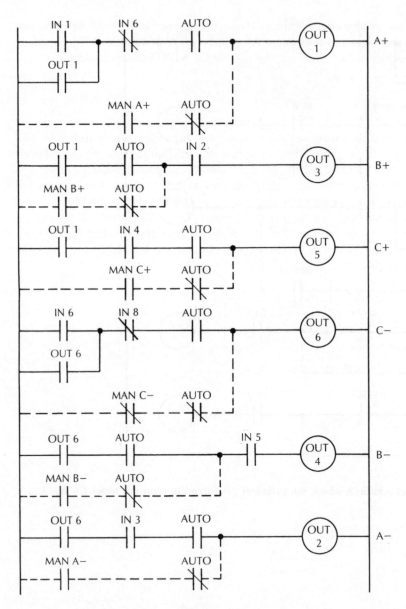

Fig. 5.12 **Manual control operations.**

forward or back in any circumstances. However, movement B+ will be prevented unless A+ is complete and movement A− will be prevented unless B− is complete. All manual operation can be seen to be disabled if AUTO mode is selected.

5.7   Summary

Timers can be used to control sequences instead of logic elements or more complex PLC ladder logic. In order to reduce the effect of timing circuits on the overall cycle time it is necessary to reduce the timer pulse duration to the shortest possible.

Many applications of machine control call for a certain number of operations and to achieve this we can use counters that can be programmed into the PLC.

Existing electronic logic systems may be converted to PLC control. Very occasionally it may be useful to convert a PLC program into its equivalent logic circuit. To achieve this conversion it is necessary to draw up a 'truth table' and to check the ladder and logic circuits.

CHAPTER SIX

Physical parts and operation of a PLC

- What does a PLC look like and how is it constructed?
- What are the basic elements of a PLC and what do they perform?
- What governs the speed, capacity and functionality of a PLC?
- What elements and features are available and how can they be used?

As described in Chapter 1, a PLC is a microcomputer designed specifically for industrial control. As such it is both physically and electrically rugged and equipped with the functionality to enable it to perform all the commonly required operations.

6.1   Basic construction

There are two families of PLC construction: the 'brick' and the 'bus'. Brick systems are designed to be low-cost solutions to small control problems. As the name infers, they are of a similar size and shape as a house brick. They are fully self contained with a power supply, inputs, output, memory and a programming port, as can be seen from Figure 6.1.

Typically they feature up to 16 inputs; 16 outputs and 1–2K of memory. They are very convenient to mount and use, simply requiring a 240 or 110 V a.c. supply and the direct connection of the input and output wires. There is, however, a drawback: they are not normally designed to be expandable. If your application, or indeed, a later modification requires more I/O lines or more memory you will need to start again with a larger unit.

The bus system takes its name from the method of data exchange within all computers, the data bus. The construction of such PLCs consists of a mechanical frame known as a 'crate' or 'rack'. Figure 6.2 shows such a system and its constituent parts. A mechanical system of guides supports a range of modules which are plugged into the rack. An electrical connector on the back of each module mates with a connector on the data bus which extends across the rack. Each of these modules

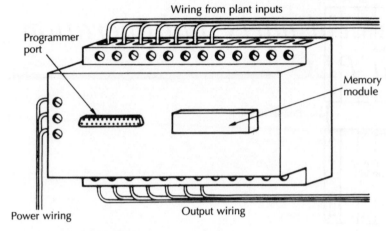

Fig. 6.1 **Brick style PLC.**

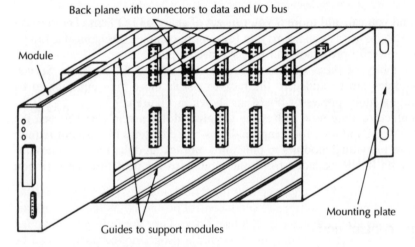

Fig. 6.2 **Typical rack style PLC.**

consists of a printed circuit board designed to perform a specific task, mounted in a protective box. The first slot is generally used to fit a power supply module which accepts 240 or 110 V a.c. and generates the various d.c. levels required by the system. The next slot is then taken by the processor card which utilizes the user's program to switch the outputs on and off to meet the changing input conditions.

The rest of the slots are available to accept whatever modules are required to match the requirements of any particular system. A number of such boards are described below. Typically only the required I/O boards and any extra memory will be installed in the rack. This means you only pay for the functionality that

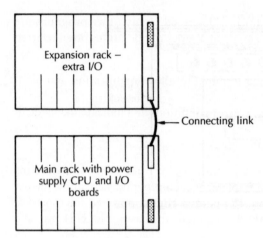

Fig. 6.3 **Typical PLC expansion rack.**

you require and you can add more if you run out of space or I/O lines. The starting price is, however, higher as you are paying for a more complex mechanical arrangement, a more powerful processor and bigger power supply.

The expandability of these systems does not stop as you fill the rack. Special expansion modules can be added to the rack to enable more than one rack to be controlled from a single processor board, as shown in Figure 6.3.

Some manufacturers are now addressing the great difference in cost between the two types of systems and are producing small bus-based systems that do not require a rack but allow individual modules to plug onto subrack modules fixed to a mounting rail. The data bus is extended from one module to the next through a ribbon cable.

6.2   Logic control unit

As previously described the central processing unit (CPU) is the 'brains' of the PLC. It consists of one or more standard or custom-designed microprocessor chips with all the support required to allow them to communicate with the programmer terminal, the inputs and outputs and memory. It is the design of this unit that determines the ultimate performance of the PLC.

The task of the logic control unit is to convert the user's program stored in memory into the control signals and commands to read the required inputs, set outputs and manipulate data.

This is generally achieved by an inbuilt program that is stored in the system read-only memory (ROM) known as the operating system. This program reads the user's program from memory, interprets what is required in each rung and calls the required system subroutines to implement it.

Apart from the interpretation program just described, many other programs are required within the operating system to carry out a series of operations often referred to as 'housekeeping'. These include:

- Communicating with the programming terminal to allow modification and monitoring of the user's program.
- Monitoring the system operation to trap error conditions and indicate the mode of failure.
- Run 'background' programs to update timers, etc.

In order to produce a PLC with increased performance it is necessary to reduce the time the system takes to understand and implement each program step, allow the use of larger user programs and access more I/O points. Higher powered microprocessors and/or purpose-designed hardware are required, increasing the cost pro-rata with the power and speed.

6.3   Memory

The memory of the CPU, as shown in Figure 6.4, can be considered to be split into a number of parts, which are used as the following functions:

- *System program.*   This is stored in read only memory (ROM) and is invisible to the user. It is normally referred to as the operating system and is responsible for carrying out the functions described in the previous section.
- *System variables.*   Again, this area is hidden from the user and consists of pointers and stored data which the operating system uses to keep track of what to do next.
- *PLC variables.*   This is a storage area where the current values of timers, counters and control relays (sometimes called flags) are located.
- *User program.* The user program (what we write) is normally stored in the PLC in an area of battery-protected random access memory (RAM). It is safe against power failures and is capable of being modified by a programming terminal as faults (or bugs) are identified. In many cases once we have proved the program we store it in a more permanent form of memory to protect it against eventual battery failure. Typically we use a programming terminal to copy the program to eraseable programmable read only memory (EPROM) or electrical eraseable programmable read only memory (EEPROM, which is more often referred to as E^2PROM). EPROM has the disadvantage of requiring a high intensity ultraviolet light source to erase its contents prior to reprogramming, but has the advantage of being four times cheaper.
- *User data storage.*   The last part of the battery-protected RAM area is set aside for us to store results and data under the control of our program. Typically we could use this area to store calibration constants or the current operating condi-

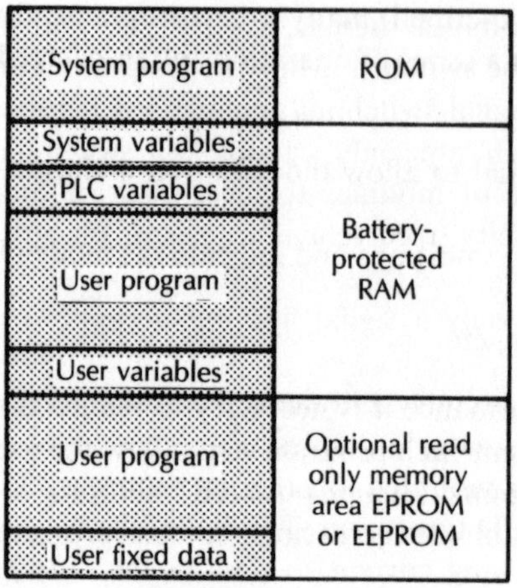

System program	ROM
System variables	Battery-protected RAM
PLC variables	
User program	
User variables	
User program	Optional read only memory area EPROM or EEPROM
User fixed data	

Fig. 6.4 **PLC memory allocation.**

tions. In cases where this data is constant we may copy some or all of it into EPROM or EEPROM to protect it from battery failure.

The memory is normally included as part of the CPU module. Additions to this often consist of plug-in modules which allow selection of the memory type and size to suit the application. In some bus-based systems extra memory is added by plugging extra cards into the rack.

6.4   Basic inputs and outputs

The input and output (I/O) section of a PLC is its connection with the real world. As discussed in Chapter 1, computer chips work with very low voltages and currents. It is only by using low voltages and low-energy signals that the required speeds and micro-miniaturization to produce integrated circuits is possible. These voltage and current levels are not, however, suitable for direct use in controlling or monitoring industrial or indeed any other kind of equipment. At low voltages and power levels electrical noise (such as can be heard on a badly suppressed car radio) would cause the spurious reading of inputs, while the power available on outputs would not even be sufficient to light an indicator lamp. Each section of the industrial and process control industries has standardized on a series of control voltages. The PLC manufacturers produce modules to meet these needs. The following list of input types is not exhaustive, but is typical.

24 V d.c.; 24–60 V d.c.; 24 V a.c.; 110 V a.c.; 240 V a.c.

Output modules also have to work at the same voltages; there is, however, the added consideration of how much power will need to be switched. Many hydraulic valves require in the order of 400 watts, while some signal-switching applications require the use of special low-noise reed relays which are only capable of switching < 100 mW. The manufacturers provide a range of modules to meet most of our needs, with relay contacts of up to 5 amp capacity, reed relays, transistor outputs and solid state a.c. relays (SCRs).

Both input and output modules require not only a signal line but a supply or common line, as shown in Figures 6.5 and 6.6.

It can be seen in Figure 6.5 that the computer circuitry and the real world voltages are normally kept separate by the use of 'Opto isolator' modules. These consist of a light-emitting diode (LED) and a phototransistor mounted together in a single package. Current flowing into the input causes the LED to turn on. This light falls on the photosensitive transistor turning it on and allowing current to flow in the output

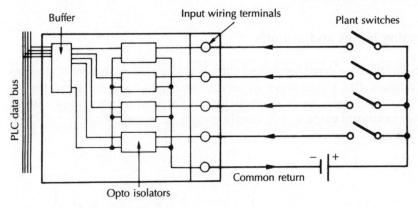

Fig. 6.5 **PLC input module circuit.**

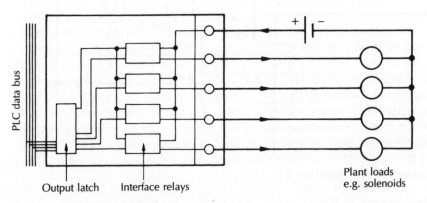

Fig. 6.6 **PLC output module.**

stage. As the only 'connection' from input to output is light, the two are totally isolated giving protection to the output from any condition on the input.

The most efficient use of space and of wire is to use one shared or 'common' supply or return wire (line) for all the inputs and another line for all the outputs on the PLC. In the majority of cases this is the preferred method to use. There are occasions, however, when this method cannot be used because the supply to all outputs cannot be connected together, or the input returns cannot be 'commoned'. To meet this need, I/O modules are available with 'isolated' inputs and output. Figures 6.7 and 6.8 describe examples for both input and output conditions where this is true.

The connections to the I/O modules are normally made through high-quality screw terminals mounted on the module. Most manufacturers allow the terminals to unplug from the module enabling it to be replaced in service without unscrewing all the terminals.

6.5   Analog inputs and outputs

So far we have only considered inputs and outputs that are the equivalent of switches and relay outputs. PLCs can, however, control and monitor much more than is possible with a simple relay control system. One of the most common requirements is to measure or control continuously variable inputs or outputs. This type of I/O is known as analog I/O.

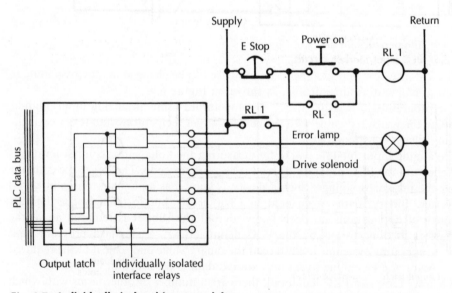

Fig. 6.7 **Individually isolated input module.**

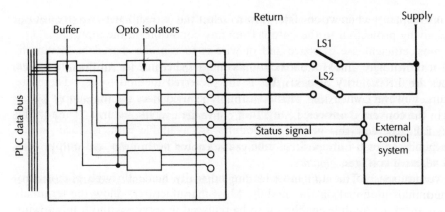

Fig. 6.8 **Individually isolated output module.**

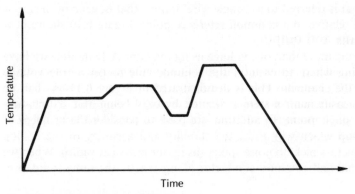

Fig. 6.9 **Temperature profile.**

For instance, an oven's temperature may need to be changed minute by minute to follow a temperature/time profile, as shown in Figure 6.9.

The program will measure the oven temperature using an analog input module, compare the measured value with the current target temperature and set or clear the heating elements as required.

The analog input module works by measuring the voltage on its input terminals and allowing the PLC program to read it as a binary code. The larger the code the larger the measured voltage. Such a device is known as an analog to digital (A/D) converter. Binary maths is covered in Chapter 10, but fortunately for the vast majority of applications no deep knowledge of the subject is required as all the hard work is done for us by the PLC manufacturer. A standard program built into the operating system is used to read the binary value and scale it to the required units. Again the technique to set up required data is described in Chapter 10.

When we look at a PLC handbook there are a number of other terms with which we need to be conversant. They are briefly listed below but are dealt with in more detail in Chapter 9 when we consider how to select the correct hardware to meet our needs.

- Full-scale voltage. This is the maximum voltage the converter can 'read'. Typical values are ±10 V, ±1 V or ±50 mV.
- Accuracy of the conversion. This is normally expressed as the number of binary bits in the converted answer, i.e. a 12-bit converter can resolve to $1/2^{12}$ or $1/4096$ of its full-scale voltage.
- The facility to read only positive voltages is known as *unipolar*, or both positive and negative voltages, *bipolar*.
- Conversion speed. The number of readings that can be converted and stored per second.

At this point we will discuss some of the common pitfalls when using analog inputs. Most analog input modules will be equipped with more than one input. The very simplest will have what is referred to as 'single wire' inputs; that is, each of the input voltages is measured relative to a common reference point. Figure 6.10 shows how this arrangement works.

If we want to measure more than one voltage using the same A/D module we have a problem determining where to connect the common line to get a true voltage measurement on all the channels. This is demonstrated in Figure 6.11. As can be seen, a single-wire measurement system is limited by only being able to measure voltage relative to a single point. In addition, we have to consider the problem of electrical noise pick-up which will affect the stability and accuracy of the analog conversion. All conductors pick up noise spikes (as in our noisy car radio). When we can run both the signal and common conductor along exactly the same route, pre-

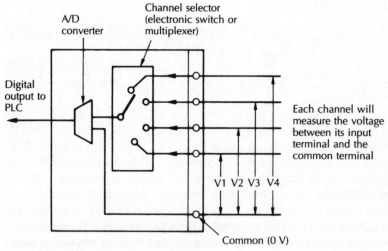

Fig. 6.10 **Single-ended A/D converter.**

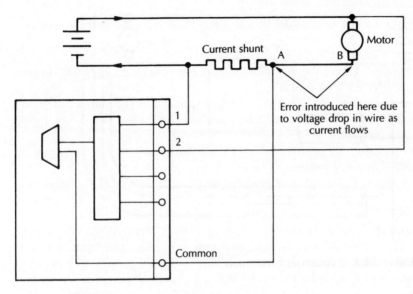

Fig. 6.11 **Single-ended A/D converter connection problem.**

ferably in a screened lead, the same noise voltage will be picked up in both leads and automatically cancelled out as the A/D converter will read the difference between the two conductors. Where more than one input is required this cannot be achieved and noise effects will become more pronounced.

We can overcome most problems by the use of 'two-wire' or 'differential' input modules. In this case each channel has its own 'common' or negative input.

We can now make connection exactly to the points to be measured, as shown in Figure 6.12.

Each channel can now read exactly the required voltage, and the leads for each channel run together in a screened cable to reduce the effects of electrical noise.

Many transducers are now available whose output consists of a current flow in the range 4–20 mA, which is proportional to the measured input; they are often termed 'process transmitters' (Figure 6.13) and have the advantage of very simple and understandable connections. In addition, current loop circuits are not easily affected by electrical noise enabling the transmitters to be sited, when required, at great distances from the PLC.

6.5.1 *Analog outputs*

Analog outputs are necessary when we require the PLC to output a variable d.c. voltage. This may be simply to drive an indicating meter or to provide a speed demand to a motor control amplifier. As with the analog input modules, the required

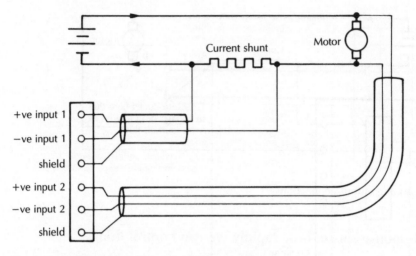

Current shunt   Motor

+ve input 1
−ve input 1
shield
+ve input 2
−ve input 2
shield

Fig. 6.12 **Differential A/D converter connection.**

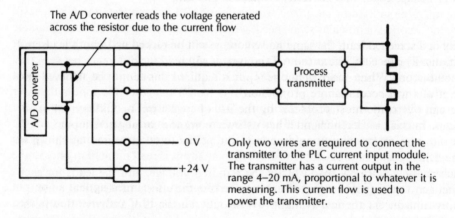

The A/D converter reads the voltage generated
across the resistor due to the current flow

Process transmitter

0 V

+24 V

Only two wires are required to connect the
transmitter to the PLC current input module.
The transmitter has a current output in the
range 4–20 mA, proportional to whatever it is
measuring. This current flow is used to
power the transmitter.

Fig. 6.13 **Current loop input**

voltage is handled within the PLC as a binary number. This is sent to an output
module which contains a digital to analog converter (D/A). As with inputs the
manufacturer will specify the resolution of the conversion accuracy and the full-
scale voltage, or current, output. Unlike the analog input there is no conversion
delay, the voltage appearing at the output immediately the binary code is output.

Most modules use a four-wire system, as shown in Figure 6.14, two wires providing the output voltage, while the other two are used as feedback into the converter to
compensate for any voltage drop in the output leads.

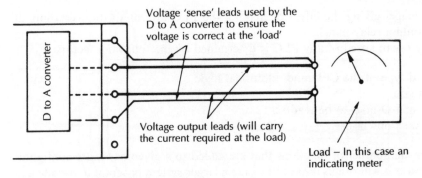

Voltage 'sense' leads used by the
D to A converter to ensure the
voltage is correct at the 'load'

Voltage output leads (will carry
the current required at the load)

Load – In this case an
indicating meter

Fig. 6.14 **Analog output connections.**

6.6 Response speed: how rapidly we can control things

Unlike a hardwired relay control system, the PLC can only do one thing at a time. Each line of our program is active in turn as the CPU interprets it. There are two important effects of this: the first concerns the consequences of the program rungs being operated in a known sequence (to be considered in some detail in Chapter 10) while the second is known as scan speed, which is the time interval between the same line of code being executed twice. In a typical system this speed will be in the range of 5–100 ms. Figure 6.15 indicates how the time is used.

In the second time window the CPU reads all the inputs into a memory area known as the 'process image'. It is from this frozen image of the inputs that our program will read the state of all inputs. This has the effect of every line of the program making its decisions on exactly the same set of input data. Were this not the case the state of any input could change during the duration of a single program scan, giving totally unpredictable results when an input is read more than once in the program.

Similarly, all outputs set and cleared as the program is scanned update an output 'process image'. At the end of the program scan all the PLC outputs are updated

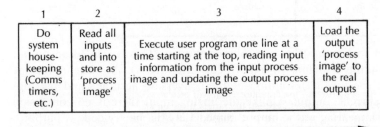

1	2	3	4
Do system house-keeping (Comms timers, etc.)	Read all inputs and into store as 'process image'	Execute user program one line at a time starting at the top, reading input information from the input process image and updating the output process image	Load the output 'process image' to the real outputs

Time into scan

Fig. 6.15 **PLC scan cycle.**

from this image, giving the effect of making the PLC appear to do everything at once, as would a relay panel.

The actual scan speed of any PLC is determined by the following factors:

- Speed and type of the CPU and additional logic
- Program size
- Number of I/O lines to be read/set
- System functions in use.

The greater the number of functions that are added to a given hardware configuration, the slower it will run as it has to execute a larger system program to decode and implement each rung of the user's program.

As we can only cause an output to change state once every scan, our response to an input changing state will at best be the time of a single scan, and at worst the time for two scans. This is shown in Figure 6.16.

If higher performance is required the most obvious answer would be to use a more powerful (faster) CPU. Such a change would, however, only give an increase in performance of 300–400 per cent. In many cases we may need improvements of orders of magnitudes. A better (and cheaper) method is to identify those few operations that require this extreme speed and handle them outside of the main PLC. Below are described several techniques and modules that can be added to rack style PLCs which introduce separate CPUs acting in parallel to handle operations autonomously. As dedicated processors they can achieve high response rates, passing the result back to the main program after they have handled the event.

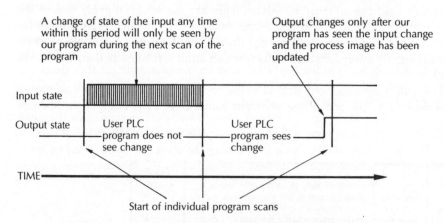

The diagram above demonstrates the conditions that will produce both the minimum and maximum delays between the program seeing an input change and responding with a change of output. Only when the input has changed before the start of the program scan will the change be seen by our program.

Fig. 6.16 **Effects of scan time on response speed.**

6.6.1   PLC interrupt facilities

The majority of medium- to high-range PLCs have what is known as an interrupt facility. This allows a limited number of special inputs to physically hold up our normal program scan and force a special section of code to be executed immediately. In this section of code we are able to directly read the 'real' PLC inputs (rather than the process image) and directly control any necessary outputs. Once the end of our interrupt code is reached, the main program will begin from where it was stopped. The time from when the input changes to when the special code takes control is known as the interrupt latency time. According to the PLC type, this is in the range 20 microseconds (μs) to 1 millisecond (ms). In some cases where we cannot use an input to generate an interrupt, an internal system timer can be set up to generate fixed short time interval interrupts forcing a fragment of our code to be run many times during the normal program scan.

6.6.2   Special function boards – to monitor and control fast events

Fast counter

As described earlier, the scan speed of a busy PLC may be up 100 ms. To count the number of repetitions of a pulse we need to read it ON and then OFF. At 100 ms per scan we could only reliably count pulses with a frequency > 5 Hz. To overcome this, special counter modules can be used which have a separate counter designed to operate up to 500 kHz. In use we preset the module by our program with a target count. When the count is reached an element in PLC memory is set (a flag) and a dedicated output within the counter module is set. Figure 6.17 shows the counter being used to count the revolutions of a shaft. Once the required count is reached the output is used to stop it. On the next PLC scan the program will see that the count has been reached and allow the next operation to proceed.

In Figure 6.17 the output line is pulsed on to call in RL1 which will start the winding motor. The relay is then latched in through an auxiliary contact on the same relay (RL1A). The counter relay will then operate when the preset number of turns is reached. This will break the latch circuit causing the motor to stop. The PLC program will detect the occurrence on its next scan and allow the subsequent operations to continue.

Position monitoring and control

An extension of the high-speed counter is the position control module. The position of shafts and machine slides can be measured using digital measuring devices known as 'encoders'. The input conditioning circuitry in these modules interprets encoder signals to automatically update the counter to represent the measured position. As with the simple counter module, our program can set up a number of preset counters, which, in this case, represent slide positions, with the outputs of these counters controlling the built-in relays. In Figure 6.18 we have set up one counter to set a slowdown relay while the second is used to control the stop relay. As all the monitoring and control are done by separate hardware, the response speed is not related to the PLC scan but solely to the speed of the counter and the output relay.

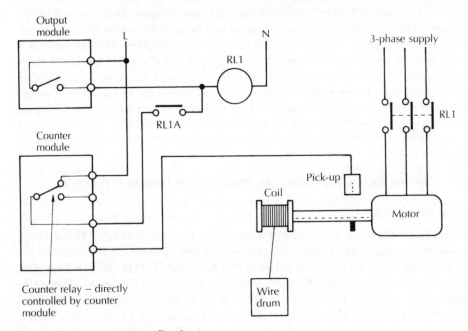

Fig. 6.17 **Fast counter application.**

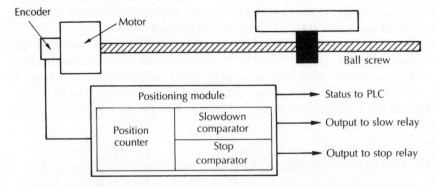

Our program will set up the slowdown and stop comparators before the move is started. The positioning module will then continuously monitor the encoder position and set the slow and stop relays as the slide moves to the required position. The module also updates the PLC memory to inform our program of progress.

Fig. 6.18 **Position control module.**

6.7 Control of continuously changing analog conditions: the PID loop

In many control situations there is a need to gradually slow down as we reach a specified target, and then maintain controllable constant conditions even though conditions outside our control are changing. There are many ways we could achieve this but PID (proportional integral differential) is a standard and well-proved technique supported by most PLCs. As an example, consider accelerating a car (automatic gearbox) to 40 mile/h in the shortest possible time, without exceeding the speed limits! As we start off we use all the power we can, perhaps easing off a little if the wheels start to spin. As we get close to the speed limit we reduce the power to avoid overshooting the target. Once at speed, if we reach a hill we need to again increase the power to maintain speed. This is a classical example of a closed loop control system where the driver uses the speedometer to compare the actual and required speeds and then makes corrections to maintain constant conditions. Figure 6.19 shows this same system in symbolic (logical) format.

The output of the comparison element is known as the error signal and is proportional to the difference between the required and actual conditions. The gain element shown is used to change the rate of response of the system. For instance, in our car example in Figure 6.19, to reach the speed limit as quickly as possible we use a high gain (foot hard down), while to improve our fuel consumption we would use a low gain and consequently take longer to reach the desired speed. There is, however, an additional problem with using high gain – we overcompensate for any varying conditions and are always 'hunting' as we keep passing through the target speed. The diagrams in Figure 6.20 show these effects. This is known as *proportional* feedback and is the 'P' part of PID.

If we reduce the gain so we do not overshoot the target we find we reach a point of equilibrium where a small error remains. This is because the error signal is insufficient to cause the change required to achieve the required condition. In our example the accelerator mechanism will exhibit friction. The force required to move it the last few hundredths of a millimetre to get to exactly the required power is greater than the tiny error signal. This can be overcome by a technique called 'error integration'.

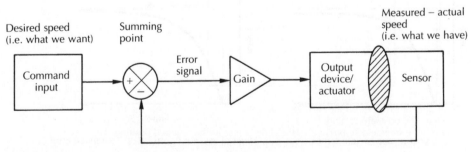

Fig. 6.19 **Logical model of closed loop control.**

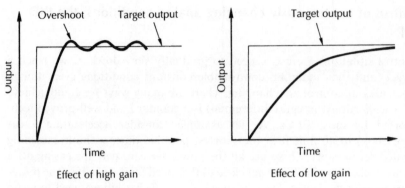

Fig. 6.20 **Closed loop gain.**

While there is any error at all an additional signal is added to it which is the product of the error and time. In our example this eventually builds up the force on the accelerator until it is sufficient to overcome the friction and make the correction.

This effect is known as *integral* feedback and is the 'I' part of PID; it is shown in Figure 6.21 and the logical model can be seen in Figure 6.22.

In our aim to eliminate, or at least reduce, overshoot the gain will have been reduced to such a point that it may take quite a long time to achieve the target. This situation can be improved by adding a further signal which is proportional to the rate of change of the error signal, i.e. as we get closer to the target we reduce the error signal. Such a system is known as a differential term; it allows us to increase the gain while reducing the risk of overshooting. Its effect is shown in Figures 6.23 and 6.24 and is the final effect known as *differential* feedback, which is the 'D' part of PID.

Such control systems using proportional integral differential controls (PID) form the basis of the majority of feedback control systems. Many PLCs have the necessary software built in or available to implement such a system using analog input and

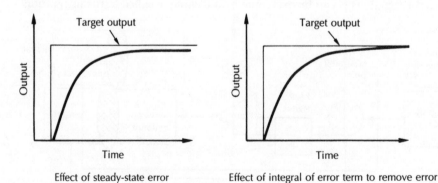

Fig. 6.21 **Elimination of steady-state error.**

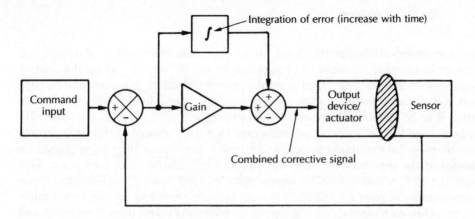

Fig. 6.22 **Logical diagram of integral of error feedback.**

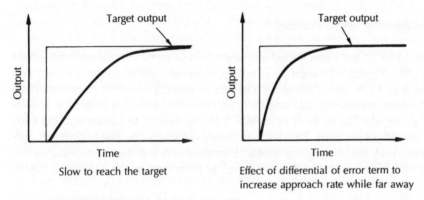

Fig. 6.23 **Effect of differential of error feedback.**

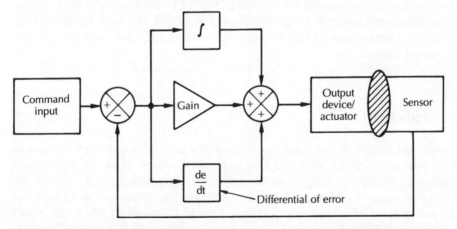

Fig. 6.24 **Logical model closed loop control with integral and derivative of error term (PID).**
(a) Schematic (b) Logical model.

output modules. Although the detail for each manufacturer will be different, the same basic information needs to be supplied by us, the user, to set up the control loop, namely: analog input, analog output, command input, gain, integration constant, differential constant, maximum and minimum output limits and how often the output is to be updated. However, we can only update once per PLC scan, the PID program itself further reducing our scan time by a few milliseconds for each control loop running. To maintain reasonable control of any system the output should be updated at the very least 10 times the rate at which changes can take place. This limits the PLC solution to slow temperature or level control applications. Some manufacturers provide a solution to this problem by providing self-contained modules with their own separate analog inputs, outputs and control processors which will control one or two feedback loops at very high speed, without any intervention or loading of the main PLC.

6.7.1   *Positioning servo control*

Servo control boards are a specialized amalgamation of servo control and high-speed counter boards. Figure 6.25 shows a simple positioning system where a motor and encoder are used to move a slide to a series of positions. The motor must accelerate up to a constant traverse speed, and then decelerate to stop the slide at a known position. Finally, the motor must provide a 'holding' torque to ensure that the slide cannot be moved off position. The diagram shows the schematic and logical arrangement required with two interacting feedback control systems: one controlling motor speed while the other controls slide position, as measured in this case by a rotary encoder.

Modules are available which contain a separate processor to implement this function without loading the main PLC processor. They are set up with the system constants using the programming terminal. The PLC program will then simply issue movement commands or trigger predefined sequences of moves. Some modules can coordinate the operation of up to three axes to allow two- and three-dimensional synchronized control.

6.8   Remote I/O

The input and output wiring costs for a physically large pice of equipment can be high both in terms of the wire and labour to install them. Most large plant and equipment is of modular design and can be wired in the same way. To enable this mid to large range, PLCs support remote I/O systems. These are small racks of inputs and outputs which we can distribute around the plant to where the inputs and outputs are physically required. The subracks are then easily connected back to the PLC using screened communication cables, as shown in Figure 6.26. We

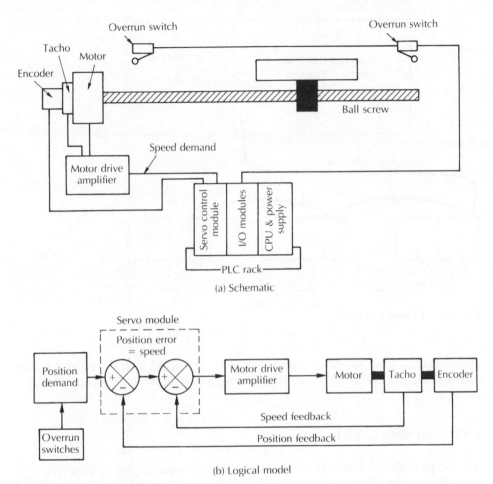

(a) Schematic

(b) Logical model

Fig. 6.25 **Servo positioning application.**

program the PLC in the usual way, a special module interpreting the I/O instructions and controlling the remote racks.

6.9   The operator interface

Communications is of increasing importance in industry. Managers want to know how many components a machine has made; the quality department will want to examine test data and set calibration constants; the operator will want to control the machine manually; the maintenance engineer will want to know why the machine has stopped. Correctly programmed, the PLC has the capability of providing this and

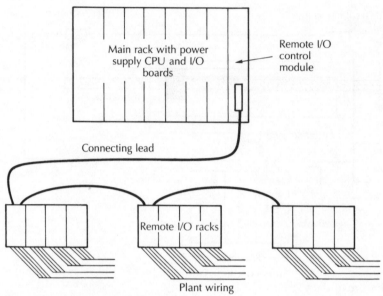

Fig. 6.26 **Remote I/O rack.**

any other data it has access to or generates. With the correct interface panel it is easy for us to display the data and receive decisions and data from the system user.

6.9.1   *Message displays*

The simplest form of operator interface is the message display (Figure 6.27). It contains an area of memory into which we can enter the text for all the required messages through a computer or keyboard. Each message is assigned a number by which it can be 'called up' by the PLC. The interface is equipped with a number of inputs which are connected to PLC outputs, these are used to select the message. Within the PLC program we can display any message by setting its assigned number as a binary code on these outputs.

6.9.2   *Operator input and display*

The two major limitations of simple message displays are that all the messages are predefined so, for example, a production count could not be part of a message and they do not include a facility to allow the operator to respond with a key press or data. These requirements can be achieved in several ways, the most common being to add an extra module to the PLC rack which communicates with a simple display

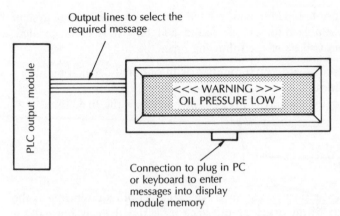

Fig. 6.27 **Simple message display controlled by PLC outputs.**

terminal, or to add a 'smart' terminal which connects to the PLC's programming port and is able to read data from the PLC's memory.

While the systems we could use will have a different method of text entry and message selection, the principles are the same for all. We will assume the operator interface selected is of the 'smart' variety and is plugged into the programming port (Figure 6.28). We start by defining and entering all the required messages into the display module through a computer or keyboard. We now have the capability of displaying not only fixed data, but of adding variable text and numbers. For instance, we could display a message which includes the production totals. This is achieved by placing special 'markers' in the fixed text where variable data is to be

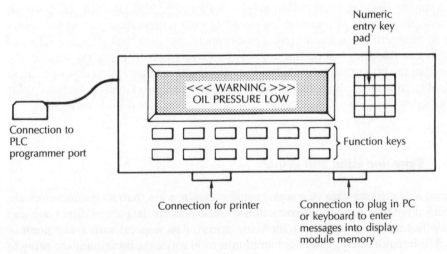

Fig. 6.28 **Operator interface panel.**

added as the message is printed. These markers tell the module's operating system where in the PLC the variable data can be found and how it is to be printed. Typically the variable data will be of the following types.

- A PLC variable displayed as a decimal number (with optional decimal point)
- Time and date information
- String of bytes to be copied directly from PLC memory into the fixed data.

6.9.3   Printed reports

The production of a printed report poses exactly the same set of requirements as the operator interface. Indeed, if an operator interface is in use there is normally a facility built in to pass selected text lines to a printer. Where no interface is in use we need to install a printer control module in the PLC. These use exactly the same principles to allow the definition of fixed and variable data already described above. In this case we enter the text into the memory of the printer control module. The printer itself is a standard device (as often used with office PCs) which is plugged into the control module.

6.9.4   Graphic displays

'A picture saves a thousand words' it is often said. One area where this is most certainly true is in the monitoring process of control equipment, where the state of many measurements and input conditions needs to be communicated with the operator. A mimic diagram, such as that shown in Figure 6.29, is ideal for this purpose. The diagram consists of a number of elements each representing part of the equipment to be controlled or monitored. These elements are then 'linked' to the values of PLC data so that they will change colour, size or value according to the value of the data. Once the programmer has set up the links the entire operation is transparent to the PLC. Control modules for graphical displays are an extension to the display modules described above, pictures as well as text being saved in its memory.

6.9.5   Time and date

In order to time and date stamp error and status reports, battery-protected clocks are built into the processors of many PLCs. These provide data in the PLC's memory from which we can read the current year, month, day, hour, minute and second to include it in any display, report or computation. A typical application is the printing of preventative maintenance prompts at predetermined times.

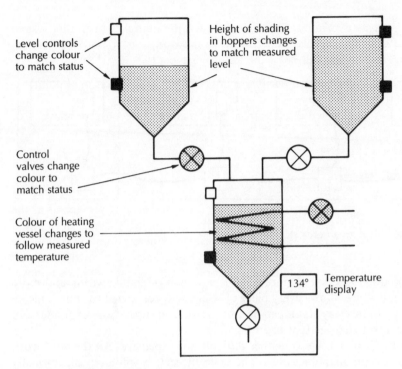

Level controls
change colour
to match status

Height of shading
in hoppers changes
to match measured
level

Control
valves change
colour to
match status

Colour of heating
vessel changes to
follow measured
temperature

134°   Temperature
display

Fig. 6.29 **Process plant 'mimic' diagram.**

6.10 'Computer' boards

Where complex data transformations or large amounts of data storage are required, a separate computer board may be plugged into the rack. It is a totally separate computer with its own memory and CPU which can run alongside the PLC processor. It can read data from the PLC's memory, perform operations upon it and return results for the PLC to handle. The module has a programming port through which we can enter and test our program (normally using the 'BASIC' language). The programming port often doubles as a printer/keyboard port allowing us to use the board as a flexible programmable data entry or reporting system designed to meet our exact requirements.

6.11 'Talking' to other PLCs and computers

There is a real need for many pieces of manufacturing or processing plant to pass and receive data not only from operator interfaces but from other (supervisory) computers. Most manufacturers build in the capability of connecting many PLCs to a local area network (LAN), through the programming port or a separate com-

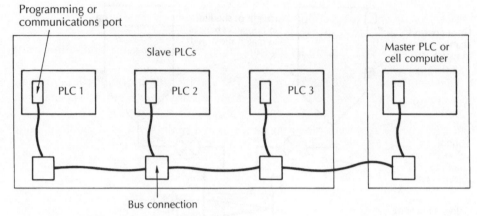

Fig. 6.30 **Simple local area network.**

munications module. A LAN (Figure 6.30) is a system of simple wiring and more complex software which allows every piece of equipment connected to communicate with every other. In this way status can be sent to a central supervisory computer and any required control changes sent to PLCs.

Connecting PLCs to a LAN is neither difficult nor expensive. All the hard work has been done by the manufacturers. The code in each is quite trivial, normally simply copying data into or out of a defined area in the PLC's memory representing nodes on the LAN. Connecting PLCs from different manufacturers is, however, a different problem. This requires specialized computers known as 'bridges' which can link one LAN to another. There are international standards that overcome this problem, the best known being the General Motors development MAP. The cost of such solutions is, however, very high and this has limited their use to date.

6.12 Summary

A PLC is a microcomputer designed specifically for industrial control applications and thus is both physically and electrically rugged in its construction.

■ There are two families of PLC construction: 'brick' and 'bus'.
■ The logic control unit or CPU is the brains of the PLC and it is this element that ultimately determines the overall performance. The input and output (I/O) modules are its connection to the real world.
■ The PLC's memory is split into four main parts which store the system program, system variables, user program and user data.
■ Where the PLC is required to measure or to output variable voltages, special analog modules can be used which convert the data handled in the PLC rungs to and from the analog voltages.

- Many other special function modules are available to handle complex or high-speed tasks such as position monitoring.
- The response speed of a PLC is the time taken between the same ladder rung being operated twice.
- This is influenced by the speed of the CPU and the length of the program. There are many ways to increase the effective scan speed without needing to change the CPU.
- A PLC can be used to display or print status and summary reports by the addition of special interface modules.

CHAPTER SEVEN

Designing and writing good PLC programs

QUESTIONS ANSWERED

- How should a program be designed?
- How should a program be tested?
- How should a program be documented?
- How can we reuse our code?

What is a good program? There may well be as many answers to that question as people answering it. An essential part of the answer must be that it performs all the functions required of it, reliably and in such a way to allow any future modifications to be made easily. One of the major problems with many large PLCs (or indeed any other program), is that a change made to overcome one problem will often cause another to arise. Such faults are often due to an obscure and expensively elusive interaction between a modification and another part of the program. Our aim is to design and write programs that do not interact in unexpected ways.

If the control task to be performed is simple, the program will be easy to design and implement and we would rightly expect the resulting program to be good. If we consider the tiny program required to turn on a lamp when two inputs are both on, the solution is simple and we can confidently expect it to work with total reliability (Figure 7.1).

We can be sure the program will work because we are able to understand all the possible conditions that can occur and check out their effects on our solution. When the problem to be solved becomes more complex, this is not the case as there is just too much to consider at once. Hence we are forced to consider, and write, one part of the program at a time. As we progress it is all too easy (in fact almost inevitable) that we forget the details and requirements of what was finished earlier, with the result that parts of the program will interact with other parts, causing the program to fail. In many situations the failures will be intermittent, only occurring some considerable time later, and requiring enormous effort to identify and correct.

The way to avoid such situations is through good and complete design before we start to write a single line of the program. We need to convert our large program into

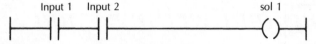

Fig. 7.1 **Example of totally reliable code.**

a number of small ones, each of which can be understood and tested separately. Once tested they can be combined, or integrated, to meet the needs of the overall requirements. The techniques of structured programming have been developed to help us in this task.

7.1   Designing the program: structured programming

The technique known generally as 'functional decomposition' was formalized by the software engineering community as an aid to the design and production of more reliable (and cheaper) programs. It is, however, equally applicable to any form of design process. The system as a whole is considered as consisting of a network of self-contained subsystems each of which has a definable job to perform. In turn, each of these can now be broken down again until each subsystem is simple enough to be very simply described and thus specified. This network of modules is shown as a structure chart, and Figure 7.2 shows how driving a car could be broken down into a number of separate modules.

By now considering and specifying the interaction of each subsystem we have a description and understanding of the entire system, but in a form that is readily usable to help us specify the program requirements. The discipline gives us a picture of all the program modules required to complete the system. In other words, we are forced to consider all aspects of the job before we physically start it. As can be imagined, there are far fewer surprises as the work progresses.

Generally the modules at the lowest level of the decomposition will contain functions performing detailed control of aspects of the plant or data transformation and storage tasks. The higher level modules will be more likely to define the operating sequence of the plant, interactions with operators, etc.

7.1.1   Definition of the function and interaction of modules

Any of our modules can now be considered as a black box that has a definable function, with inputs and outputs, that consists of its interaction with the PLC I/O or other software modules. For example, our navigation module can be defined as being required to determine and update at all times the correct road to achieve the final aim of arriving at our destination. To achieve this it will require a number of inputs which can be described in words but are often more clearly defined in a simple diagram, as shown in Figure 7.3.

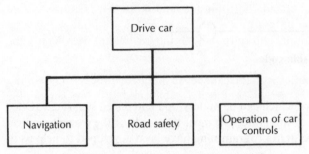

Each of these modules can again be broken down until the functions within it are easily defined and can be written as a self-contained block of codes

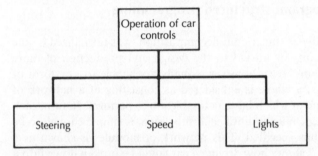

Fig. 7.2 **Functional decomposition of car driving example.**

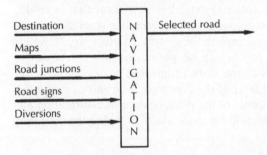

Fig. 7.3 **Simple input-output diagram of a software module.**

The modules we define must not only be simple, but the links with all other modules must be minimized and easily definable. These links may be data passing within the PLC, data links to other PLCs or computers, or conventional inputs and outputs to the plant to be controlled.

1 By defining the operations within the modules we define the software we need to write.

2. By defining the modules' data and control inputs and outputs we define the data and control flow within the program.

3 By defining the modules' inputs and outputs with the 'real world' we identify the PLC I/O requirements.

The decomposition of a complex system can be an iterative process. Our initial decomposition may produce a set of simple definable modules, but should the links between them be complex we will need to reconsider the module boundaries. If we consider our example of the car driver, it would be tempting to define one module to control the operation of the clutch and a second to control the operation of the gear stick. This would, however, create a pair of modules that need to 'understand' each other's operation if we are to prevent movement of the gear stick while the clutch is not released! This relationship would be apparent from a number of inputs to both modules appearing as the outputs of the other. A better design would be for us to design a single transmission control module to calculate when gear changes are required, and control the sequence of both the clutch and gear stick. We could now use separate modules for the control of both clutch and gear stick, but they would simply respond to the logic of the transmission control module.

7.2   Implementing the solution: writing and testing the code

Just how our set of modules are organized is somewhat dependent on the PLC to be used. The principle is for each module to remain dormant, waiting to perform its task. This can be achieved by various methods according to the PLC to be used, but will involve the skipping or disabling of the blocks of code when the functions are not required. A number of methods to achieve this are discussed and demonstrated in the advanced programming discussion in Chapter 10.

7.2.1   *Module write and test*

Each of our modules is now well defined and should be no larger than can be written and, whenever practicable, tested in the same day. There is considerable debate as to the correct order to write and test a set of modules. In general, the best approach is to start with the modules that can easily be tested in isolation. Once all these are written and tested they can be left in place in the PLC. The modules that rely on them can now be written and tested utilizing our now proved modules. The testing process must ensure that the module will work correctly in all possible situations. Chapter 8 has further information on how the programming terminal can be used to control the program operation.

We must avoid the trap of simply testing to prove that our modules work! In addition, we must prove that they will not fail or produce incorrect results if unexpected conditions are encountered. To achieve this the following list of tests should be carried out (where applicable):

1   All input variables must be varied throughout and beyond their working range to ensure that the outputs work correctly. It is not always practical to perform these tests on every possible value, but experience shows the following to be essential:

0	+1	−1
Full-range value	Full-range value + 1	Full-range value − 1,
Low-range value	Low-range value + 1	Low-range value − 1.

Where there is more than one variable to input if they interact in the module in any way, the tests should normally be performed with all combinations of inputs in each state. The test data used should be recorded to allow retesting of the module should any changes be made.

2   Where a module has inputs or flags which control its logic, all combinations of lines must be set to ensure that it performs in the expected way.

3   In the previous tests the modules can be looked upon as a 'black box', simply setting inputs and checking outputs. Some modules will, however, contain logic where the order in which inputs are changed will affect the operation of the code. To carry out tests, the programming terminal is used to examine the code of the module as its inputs are changed in the order required to test the operation of each section of the logic. A typical example of this is where the program is controlling a process that is sequential. In such a situation we need to keep careful track of the progress of the code and test all the effects of inputs appearing in both the expected and unexpected orders. In order to test all the possible outputs that the code could produce, we need to produce a test list as we design and write the code to enable us to force (see section 8.3) all the error conditions designed into the program.

7.2.2   *Module integration*

Once we have finished the base level modules, others that use them can be written and tested in the same way. We can leave the finished modules in the PLC to scan for outputs from the module under test and return any results required. In this way we build the entire program module by module, allowing us to concentrate on testing one small aspect of the program at a time. The final result is a well-tested, well-structured program that is easy to understand and modify in the future. Modifications to the code can be carried out with confidence as we understand and have proved all the interactions with the rest of the modules, eliminating the causes of unexpected interactions.

7.3   Simulation

As already described in section 4.1, it is not always possible, practicable or desirable to test a program with the real hardware to be controlled. In such cases we can use input simulation or switch boxes connected to our PLC to set up the conditions required to test our program. For the vast majority of cases this is the preferred testing method for all the lower level modules where we can easily force all the possible conditions to enable us to test the accuracy of the code. If we use the actual hardware, it is too easy to fall into the trap of letting the hardware cycle through its sequence of operations, only testing the code to 'prove it works'.

In addition to simple switch boxes, a panel containing adjustable voltage sources to force analog inputs, and voltmeters to monitor analog outputs, is invaluable when the program has to monitor and/or control analog variables.

Writing our program in modules makes it easy to substitute a special test code in the place of one or more modules. For example, this technique can be used to great effect to simulate the operation of a section of the hardware to which it will be connected. The module that controls the function is replaced (only after it has been fully tested!) by a test code obeying the input and output rules. The rest of the modules can now be tested without the need to continuously operate switches to simulate the detailed operation of that section of the hardware.

7.4   Program documentation

The documentation of PLC software is often poor. Many programmers seem to believe that a listing of the program code is sufficient. While it may be good enough for them while the program is still fresh in their minds it would take someone else a considerable time to be able to maintain it when the need arises. Documentation for a PLC project is not just a record of what was done but can also be a powerful design and maintenance tool. It needs to address all the areas described in the following sections.

7.4.1   Project definition and documentation

Every program requires an overall description of what it is designed to achieve and how it is structured to meet these ends. The first of these needs is normally met by the requirements specification for the system. This must be full and detailed enough to let us design our program to meet the needs of the overall project. It needs to define every operation the program will need to perform. The description of the program structure may be a written description of what operations each section of the code will perform and how they interact. If, however, a functional decomposition has been carried out, the resulting structure and I/O charts meet these requirements excellently.

An example of a simple specification for a project and the top level functional decomposition is included at the end of Chapter 10.

7.4.2 Module description

Each module must have a written description of the operations it performs, the full list of its inputs and outputs and any assumptions made. This can be a separate document to the program, but where the programming software allows, it can be conveniently included as a part of the program documentation. Figure 7.4 shows the documentation page of a typical program module.

7.4.3 Symbols

If we allocate a meaningful symbolic name to each input, output and element of data its appearance in the code can be more readily understood. For example, the function of inputs named START and STOP will be immediately apparent without the need to refer to a drawing of the machine connections. Such names and descriptions are assigned to inputs, outputs and data using the symbols feature of the programming software. An added advantage is that we produce a well-documented list helping us to avoid the common mistake of accidentally re-using an area that is already in use.

```
================================================================

Title        Display message on remote text display
Written By   Bill J.
Date         29.1.92
Version      V 1.1

Mods         V 1.0  29/1/92 Original release
             V 1.1  12/2/92
             Only update if disable flag is clear (inhibit overwriting of fatal
             error messages)

Description
Sets a message number into the display data block and sets the new message flag
to initiate a message update.

Data input
Input data constant in decimal format range 1–1000 Message numbers outside
this range are ignored

Test data
0, 1, −1, 999, 1000, 1001, 499, 500, 501
================================================================
```

Fig. 7.4 **Typical module description block.**

Table 7.1   **I/O symbols definition**

Operand	Symbol	Comment
IN 32	CLAMP	Clamp Ip's
IN 33	UN-CLAMP	Un-clamp Ip's
IN 34	COMMONPB	Common PB for clamp/un-clamp
IN 35	LS2	Contacts are back
IN 36	I FAIL 1	Current fail low level
IN 37	I FAIL 2	Current fail high level
IN 38	DB FAIL1	Noise fail low level
IN 39	DB FAIL2	Noise fail high level
OUT 1	PASS LMP	Pass lamp
OUT 2	F SOL	Fail sol lamp
OUT 3	F CURRENT	Fail amps test
OUT 4	F NOISE	Fail noise test
OUT 5	F SPEED	Fail speed test
OUT 6	RESET	Reset the counter interface
OUT 7	CLAMP	Clamp the starter
OUT 8	UN-CLAMP	Un-clamp starter

The fragment shown in Table 7.1 is taken from an input allocation list demonstrating how each I/O point is defined and described. In this table the OPERAND defines the PLC inputs. SYMBOL is the description that will appear with the contact in the code while COMMENT more fully describes the function of the line. Note that the selection of inputs and outputs is not in itself significant.

7.4.4   Code comments

While the module definition describes the operation performed by a software module, the code comments describe how the operation is achieved. Such comments can be entered for each line of a program to describe its operation.

The segment of code presented in Figure 7.5, which controls the movement of a pneumatic cylinder, shows how the use of both symbols and comment with each line of code makes the operation of each line of the program easy to understand.

7.4.5   Revision history

After a number of modifications have been made to a program, perhaps by different people, it can be difficult for us to identify what has been changed and why. The simple practice of allocating a version number to each program module, and to the program as a whole, can help to alleviate this problem. The module description shown in section 7.4.2 (Figure 7.4) can be seen to contain a version number and a

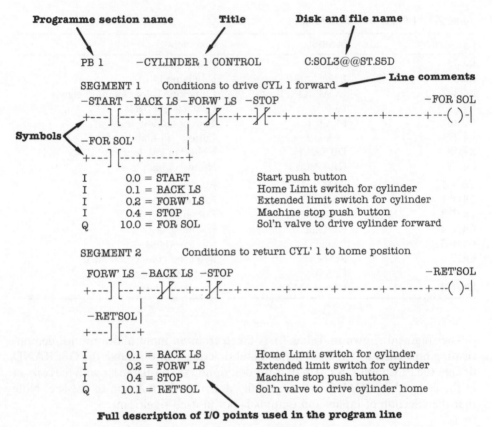

Fig. 7.5 **Illustration of good commenting practice. Reproduced with permission of Siemens.**

description of any changes. The example in Figure 7.6 shows the version control sheet for the entire program, which lists the changes to the program as a whole and the modules modified to achieve it.

We should always enter the version number of each module and the overall program into the data area of the PLC itself, enabling us to determine at any time exactly which version of the program is currently in memory.

7.5   Motion and phase control diagrams

These diagrams perform two functions. Firstly, they check the validity of the ladder diagram logic, e.g. for hidden locked signals, failure to turn off some device within the machine, having switched it on at some part of the cycle of operations, etc. Secondly, they are useful in trouble-shooting maintenance operations since the status

REVISION CONTROL SHEET – STARTER MOTOR TEST

FILE   STARTER.S5
DATE   13/4/91
AUTHOR   W. JEFFCOAT

VERSION   V 1.0
CHANGE   Original release
MODULES CHANGED

VERSION   V 1.1
CHANGE   Commissioning found need to add delay in pull in test code
MODULES CHANGED   PB 10   V 1.1,   PB 100 V 1. 1

VERSION   V 2.0
CHANGE   Addition of speed measurement during run test
MODULES CHANGED   PB 20   V 1.0,   PB 100 V 1.2,   FB 40 V
1.0

Fig. 7.6 **Revision history.**

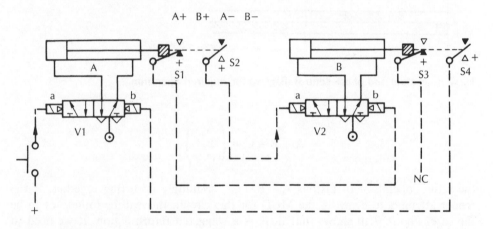

Fig. 7.7 **Required circuit.**

of every input and output is known throughout the cycle, relative to time. (This aspect of the diagrams is described in section 11.8.)

The method of construction of the diagram is best illustrated by examples.

Consider A+ B+ A− B− as shown in Figure 7.7. Firstly, the MCD format is drawn as shown in Figure 7.8 and by reference to the operational sequence and wiring details a total 'picture' of the operation is built up relative to time, SW2 provides a signal to solenoid V2 which then operates piston B+. This cause–effect action is repeated for all other parts of the system, resulting in a complete diagram.

Note that a MCD can be drawn up for any control system logic, relay, PLC, etc. Example 2, shown in Figure 7.9, illustrates a circuit which at first sight works as follows:

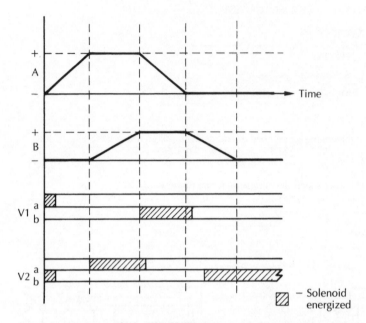

Fig. 7.8 **Motion and phase control diagram for required operation.**

$$A + B + C + \begin{pmatrix} A- \\ B- \\ C- \end{pmatrix}$$

the return operation of pistons A– B– C– seemingly occurring together. If the reader attempts to draw up the MCD for this circuit, this will be found not to be the case. Figure 7.10 shows that there is a staggered return action. If we need to return all pistons simultaneously this circuit will not accomplish this task. The MCD in this case is being used to check the operational sequence against the specification and shows very clearly that the desired simultaneous return of ABC is not being met.

7.6   Program libraries

A further advantage of writing software in well-defined modules is the possibility of re-using software from previous programs. We will quickly find that certain modules will have application over and over again. Such modules can easily be listed in a library list to keep them in mind as we design the structure of future programs. Re-using such pre-proven code can significantly reduce the time and cost of writing software.

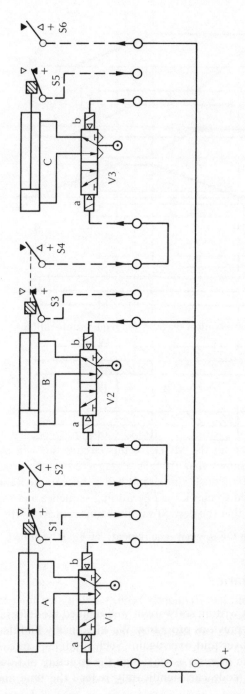

Fig. 7.9 **Connection circuit to meet needs.**

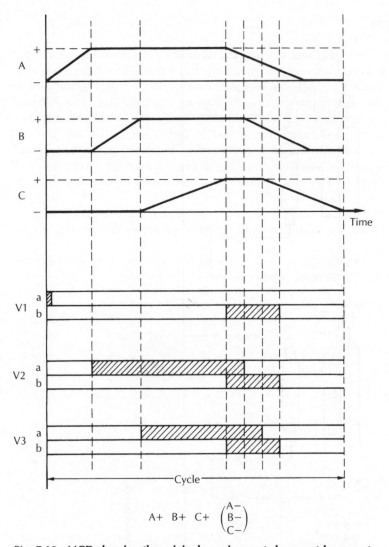

Fig. 7.10 **MCD showing the original requirements have not been met.**

The concept can be, and has frequently been, taken a step further in cases where a new project can be seen to have a similar function and structure to an existing program. In such cases we can use the framework of the high-level, as well as the low-level, modules to ease both the design and production of the program.

The code originally written for a d.c. motor test was reworked to produce a compressor test (and subsequently two further tests). Over 50 per cent of the original code was re-used allowing a considerable saving in time and cost.

7.7   Summary

A good program is one that will work as intended in all circumstances. Such programs only occur if they are well designed and well written. Designing good programs is not a matter of luck but involves careful and detailed planning and implementation. We can design our overall program using the rules of 'structured programming'. The implementation should then follow the sequence: individual module write and test; module integration; system simulation; system test.

The project documentation should be produced as the project proceeds. It is used to record, validate and track modifications to the design, and must cover the overall and detailed design of the program. Symbols and comments should be used throughout to help explain the code.

The production of phase control diagrams to describe the operation of ladder rungs is a powerful method of documentation and of validating the program against the requirements.

Program modules that can be re-used can be placed in documented libraries to be available for re-use in future programs.

CHAPTER EIGHT

Programming hardware

- How can we input our program?
- How can we influence the operation of our PLC?
- How can we document our program?
- How can we store our program?
- What kind of programming terminals are available and how do they compare?

To write or test any kind of PLC we need a programming terminal or 'programmer' that is matched to the PLC we wish to use. They are available in many forms, from a hand-held terminal costing one or two hundred pounds to a multi-tasking multi-user computer serving several programmer stations and costing many thousands of pounds. It is the basic tool kit for the PLC programmer and maintenance engineer. It enables us to write, document and copy the program and then 'look' inside the PLC to monitor execution. The aim of this chapter is simply to give an understanding of why and when we might use the features available to us. The detail of how each is performed varies so dramatically from manufacturer to manufacturer that operation details are best left to the operating manuals.

8.1   Programming

Our program is entered into the PLC through the keyboard of the terminal. The majority of programs are written and entered as a series of ladder rungs. This is not, however, the only method of representing a program. Two others in common use are:

1 Statement format, where the program is entered as lines of instructions, much in the way a conventional computer program would be written.

2 Control flowchart, which is a graphical technique representing the program logic as a series of 'logic' blocks.

Figure 8.1 indicates how the same program would be represented in all three formats for a Siemens PLC.

The program to allow the entry of code in statement format is relatively simple, allowing us to type the instructions for each line of the program, and then check its

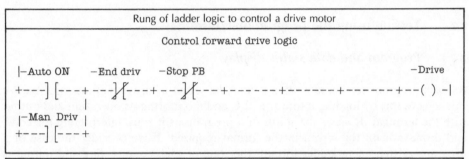

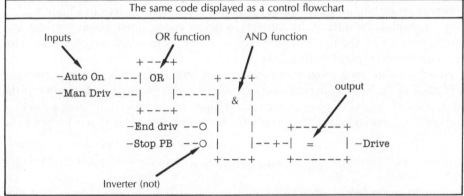

As each line of the program is executed its result is stored in the Result of Logic Operation flag (RLO). This is then combined with the input in the next line according to its logical operator (AND, OR, AND NOT, OR NOT) until we reach an output instruction (=) where the output is set to match the RLO.

Fig. 8.1 **Comparison of the three common program representations. Reproduced with permission of Siemens.**

validity normally on a line-by-line basis. The other two formats require a more complex program that will allow us to draw and modify our graphical representation of the program on the screen while storing it in the form required by the PLC. This is achieved by the use of specifically defined and identified 'function' keys on our keyboard, allowing movement around the screen and the entry of the contacts, coils, timers, etc.

8.2   Testing using the programmer/terminal

8.2.1   *Program and data status display*

As each program section is written it must be tested to prove its functionality. We can achieve this by loading it into the PLC and monitoring its operation and output with the terminal. It views the status of a program as it runs, interpreting the code, and displays it on the screen in the format required. If we consider the case of a program written using ladder format, the state of each contact, flag and output is indicated by highlighting the 'Made' conditions. This can be seen in Figure 8.2.

By examining the state of the contacts in ladder rungs, timer, counter values, etc. we can clearly view the status of each section of our program, identifying reasons for program errors or plant malfunction.

However, while such program views are a powerful tool to monitor operation and help identify faults in individual sections of the code, finding problems due to the interaction of several sections of the code is much more difficult, involving tracking and searching for code that can modify the output or data value that is known to be

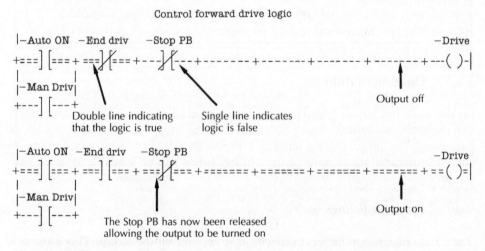

Fig. 8.2 **Ladder rung status display. Reproduced with permission of Siemens.**

at fault. In such cases we also need to understand the overall structure of the code. This is where good initial design (see Chapter 7) and good documentation prove essential.

We can also monitor data through our programming terminal. By selecting the area we wish to view the current value of data is presented on the screen, the format of each element being selectable to match its use in the program. Typically, but not exclusively, we could choose binary, decimal or hexadecimal format, or indeed view the data as text.

8.3   I/O and data forcing

Certain fault conditions' program and I/O testing require the monitoring and 'forcing' of inputs, outputs, flags and data areas. This facility built into programming terminals enables us to control the state of the selected entity, overriding the operation of the PLC inputs and/or program. The technique varies from manufacturer to manufacturer, but in general terms forcing operations are an extension of the status monitoring function. We first select the input, output, flag, data, etc., through the appropriate status display and then select it by positioning the cursor over its image on the screen. Through the use of a dedicated function key we can now force it on or off. In the case of data, the value can be modified by over-typing a new value.

This is a very powerful feature and we must use it with extreme care. It is all too easy to force an output that may cause damage or even injury by moving part of the hardware unexpectedly. Some common applications of the technique are

- exercising outputs to ensure that each is correctly connected to the hardware
- forcing inputs and flag values to check that our code responds correctly to fault conditions, or to simulate the operation of hardware not yet connected
- modifying the data values input to a program module to test its operation.

8.4   Documentation

As discussed in Chapter 7, good program documentation is essential to enable others (or ourselves two months later!) to understand the overall design and operation of programs. The programming terminal's functions will vary from manufacturer to manufacturer, but all or most of the features below will be available to us.

8.4.1   *Program listings*

Each rung of a program is printed out as it appears on the screen. This gives us a written record to allow reference to sections of the code other than that displayed on the programmer screen at any time. This listing should contain not only the basic code, but also the following information, which has been fully described in Chapter 7:

■ A list of all the inputs, outputs, flags, timers, counters, and data used
■ The symbolic names for all I/O and flags as an aid towards understanding the program
■ A description of the function of every module or section of the program
■ Detailed line-by-line comments of the operation of the program
■ A history of the modifications to the program.

8.4.2   *Cross-reference*

Even in well-designed programs we can make mistakes. The situation may occur when an output or flag is unexpectedly changed by part of our program. This is commonly due to our having failed to keep a good record of flag usage, part of the program being scanned when it should not be, or simply by bad design! A cross-reference program is normally available to help us find these faults. This will produce a list of every flag, input, output, timer, program module, etc., and list where it is used in the program.

8.5   Program storage

The program in a PLC is often stored in battery-protected RAM, which presents the danger of data loss if the battery is not changed at regular intervals. This can be overcome by using the programming terminal to copy the program into a non-volatile memory module (normally EPROM), which is then plugged into the PLC.

The program we enter and store in the programming terminal, along with all its documentation, is known as the 'source code'. In most cases this will be stored on the hard disk. To produce back-up copies and long-term storage we can then transfer it to floppy disks, or data tapes. There are, however, still a few older systems in existence that depend on audio cassettes for program storage. Whichever system is in use, however, we must always make a duplicate of the program and keep it separate from the master copy.

If we are faced with the need to work on a program for which we do not have the original source files, it is possible to read the code directly into the programmer from the PLC. In such cases, however, comments and symbols will not be available.

Fig. 8.3 **Typical hand-held terminal.**

8.6 Types of terminals

8.6.1 Hand-held

The cheapest programming terminal is the hand-held type, looking rather like an overgrown calculator and costing a few hundred pounds. It simply plugs into the programmer port of the PLC from which it obtains its power. Such devices are capable of allowing us to enter a program directly into the PLC, copy the PLC contents into EPROM and can be used for testing programs and the system. They do not, however, have any form of disk storage, preventing the entry of comments and symbols. They are useful as an easily carried tool for post-commissioning investigation into control systems problems. A typical terminal of this type is shown in Figure 8.3.

8.6.2 Desk and laptop

All serious programming and development is carried out using a full function programming terminal. Each supplier sells a range of such terminals (costing from £2500 upwards) with increasing speed, memory, disk space and screen resolution. They come ready configured with all the required software to develop PLC programs. Such systems are, of course, expandable, with extra packages available to allow specialized hardware to be configured and techniques such as Grafset or EDDI to be used (Chapter 10).

Physically the hardware is a specially ruggedized computer in either desk or laptop configuration. In addition to the normally expected facilities of disks, serial and printer ports, it is provided with a special communications link to suit the PLC programmer port requirements and an EPROM programming facility.

8.7   IBM PC-based solutions

The vast majority of the computers described above are based on the IBM PC (Figure 8.4). It is possible to buy and install the required software on an existing PC. There are a number of minor disadvantages to this approach: (1) as the function keys are used as the special 'ladder rung' keys, a chart of the key positions is required; (2) separate equipment is required to connect the PC to the PLC, and to program EPROMS.

Choosing to use an existing PC opens up the option of buying your programming software from third-party software houses. Such software is normally cheaper than the offerings of PLC suppliers, and in some cases is easier to use. This can, however, leave you without support if your program fails to execute correctly, and although such instances are rare, they can and do occur.

If you use PLCs from more than one manufacturer a standard PC is probably the best general solution. The same machine can then be used for all applications, saving the purchase cost of individual equipment from each supplier.

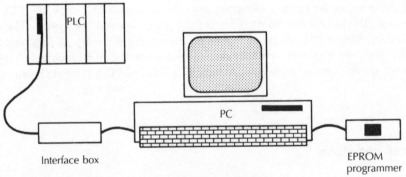

Fig. 8.4 **IBM PC used as a programming terminal.**

8.8　Summary

To program or test any PLC we need a programming terminal or 'programmer' which is matched to the PLC. This comes in many forms, ranging from a hand-held unit which looks somewhat like a calculator to desk and laptop terminals. It is also possible to program many PLCs using an IBM personal computer running an appropriate software package.

Terminals are used for programming, testing programs and I/O status and data forcing. The program of a PLC can be stored in battery-protected memory or in more secure memory, normally an EPROM.

CHAPTER NINE

Selecting a PLC for the application

- How do we understand and estimate the requirements for a PLC?
- How do we select the I/O hardware?
- How are I/O circuits configured?
- How do we size the processor and memory for our application?
- How do we select a supplier?

The choice of a PLC for a particular application can be bewildering. The range of suppliers is vast, many offering a number of alternative product ranges, with any number of modules to perform a range of esoteric functions. Our choice must meet the job and customers' requirements, provide extra capacity to enable future modifications and provide an acceptable cost solution.

We have to make choices balancing the cost of extra, or more expensive hardware against the time required to program algorithms that allow us to use cheaper hardware to meet the system requirements. Each case has to be considered on its merits. Beware of the common trap of underestimating the time taken to write such code!

9.1 Estimating requirements

The starting point in determining any solution must be to understand what is to be achieved. In an ideal world our customer (even if we are building a system for ourselves) will have produced a detailed specification of the requirements. If this is not the case, we must start by preparing one.

9.1.1 System definition

In Chapter 7 we discussed program design, breaking down the task into a number of simple understandable elements, each of which can be easily described. The same technique of functional decomposition is equally applicable to defining the whole system, both hardware and software, as it is in defining the program alone. The most common mistake is to attempt to handle the entire system as one unit. When such an approach is made we will immediately select solutions for the parts of the system we 'know' are going to be a problem, or the parts we immediately 'know' how to solve. This approach diverts the design and equipment selection away from what is required to solve the real problems, and leaves us with a solution that may be far from ideal.

A worked example can be found in the Appendix, which shows a typical decomposition of a fairly complex application and sample I/O diagrams.

9.2   Choosing the correct I/O hardware

With an understanding of the entire system we can start to estimate the PLC requirements. For each module the inputs and outputs can be categorized for type and speed of operation. Section 6.4 described the various types of input and output modules but here we will consider the selection criteria.

By knowing the number of any type of I/O lines we need and the number of lines available on a given module, the final shopping list of modules and the size of the PLC system are determined. In addition, build in at least 20 per cent extra capacity to allow for future modifications or to solve problems identified during commissioning.

9.2.1   *Simple I/O timing considerations*

For every element, we need to determine how fast the subsystem of input program and output must react to changing input conditions. The speed of operation will be the sum of the input hardware delays *plus* the PLC scan time *plus* any output hardware delays. In the vast majority of cases a time delay of 100 ms or greater is not significant. Typical instances in which this may not be the case are pulse counters, or where a movement has to be stopped in mid-stroke. To determine the required response speed, we need to consider each of our defined modules with its inputs and outputs. The effect of control decisions being taken at various rates can then be considered and the slowest rate determined. For example, Figure 9.1 shows a simple tank level control application. If the flow rate is known to be 10 litre/s and we want to maintain the volume of liquid in the tank to ±1 litre we need to read the level input and make a decision to set or clear the flow valve output, at the very slowest, every 0.1 s. This can be determined by calculating the time it takes for our minimum control quantity to flow into the tank, i.e. 1 litre ÷ 10 litre/s.

In the second example (Figure 9.2) we need to stop the cylinder mid-stroke to an accuracy of ±0.2 mm. We know that its maximum speed is 100 mm/s, so to achieve

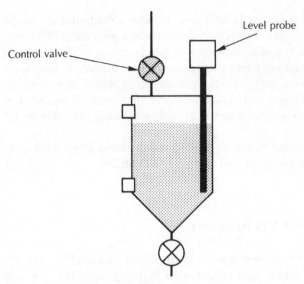

Control valve

Level probe

Fig. 9.1 **Tank level control.**

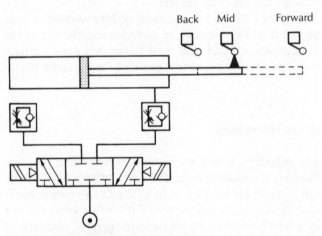

Back   Mid   Forward

Fig. 9.2 **Controlling pneumatic cylinder.**

this we would need to be able to make a decision to set or clear the control valve every 2 ms, i.e. the time taken for it to move 0.2 mm ÷ 100 mm/s.

To achieve this we would require a scan speed that would be difficult to guarantee except in the smallest programs using a fast PLC. An interrupting input is indicated. The switching speed involved will probably also cause a problem because of delays in the electronic and pneumatic hardware. This will require us to stop the movement a known distance before the target position. Figure 9.3 shows how these delays are introduced.

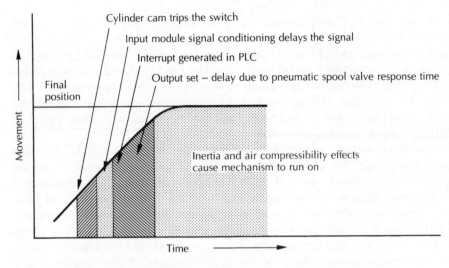

Fig. 9.3 **Time delay effects when stopping a pneumatic cylinder.**

This, of course, assumes that the cylinder is moving at constant speed as it trips the switch and that all the electronic and pneumatic delays are constant. If d.c. inputs and outputs are used, this is a reasonable assumption. When a.c. I/O lines are in use there is always an additional 10 ms uncertainty, as described below. The more normal action would be to use a second switch to slow the cylinder down before it reaches its stop position and then use the final switch to stop it on station.

Input selection

For each input we need to determine the following points:

1 *Voltage level*: most systems in the UK use 24 V d.c. or 110 V a.c. for inputs and/or outputs. 24 V is rapidly becoming more favoured, particularly for input circuits using solid-state proximity switches.
2 *Response speed*: d.c. input modules typically have a response speed in the range 2–5 ms. With a.c. inputs a slow filter circuit (> 10 ms) has to be fitted within the input board to prevent an input which is ON appearing to go OFF every time the current reverses. This sets the response to 10–20 ms (based on a mains frequency of 50 Hz).

Outputs

For each output we need to determine the following points:

1 Voltage level: the considerations are the same as for input circuits.

2 The power that PLC outputs need to switch varies greatly. A lamp may only require 5 W while a hydraulic solenoid can draw up to 400 W.
3 Output resistance and electrical noise can be an issue in cases where low level signals are to be switched. For example, consider the case where a number of low level (< 100 mV) analog voltages are to be switched into an analog input (a multiplexor). In such a case, voltage drops or induced voltages across contacts designed to carry high current are not acceptable, and modules specified as using 'signal' or 'mercury wetted' relays are required.
4 The use of a.c. outputs can often be an advantage. In most cases the voltage is higher (commonly 110 V a.c.), giving a fourfold reduction in current for any particular load, when compared to 24 V outputs, the consequent reduction in the wire size required giving a reduction in wiring costs. A second and often more important advantage is the reduction of electromagnetic interference (EMI). When a contact opens or closes the current will attempt to flow across the momentary gap between contacts. The consequence is a spark of short duration which generates radio waves at very high frequencies. These are picked up in all the wires of the control panel. A PLC is designed to cope with such interference and should not misbehave. However, not all equipment is so robust, so a.c. solid-state outputs are designed to switch on and off in such a way as to generate an absolute minimum of EMI.

9.2.2   Analog I/O modules

When we set out to select analog modules there is a need to understand a number of terms used to describe their performance; this allows us to match hardware performance to requirements. This section will describe those terms that are common to both input and output modules.

■ 'Resolution' defines how accurately the analog to digital (A/D) or digital to analog (D/A) converter within the module can represent an analog voltage as a binary number, or vice versa. For instance, a 12-bit converter can resolve its full input voltage into 4096 intervals. (Binary numbers are more fully described in section 10.1.3.) Such a converter with a full range input of 10 V can measure the input voltage to ±2.44 mV, i.e 10 V ÷ 4096.

  If the converter is bipolar with the specified 12 bits, the voltage range is ±10 V. A 20 V range, of course, would give a resolution of 4.88 mV. Such converters are often described as 11 bit + sign. The resolution must be chosen to match the application's requirement. If a voltage has to be measured with an accuracy of ±5 mV, we should aim to have at least twice that resolution in the converter to allow for temperature and time drift as well as all the other inaccuracies that reduce the measuring accuracy from the nominal A/D or D/A converter specification.

- 'Isolation' refers to the ability of each input or output to work at voltage levels independent of the system ground. This concept and its applications were covered in section 6.5.

Inputs

As discussed in section 6.5.1, to select an analog input module the following points need to be considered:

- **Voltage level** The maximum voltage of the input to be measured must be determined. A module with a maximum range just greater than this level would normally be the best alternative. If, however, there are a number of analog inputs to be measured at differing maximum voltages, we must either use a module to cope with the highest of the voltages or use a lower voltage module and resistors as potential dividers to normalize all inputs to the same range. If any of the voltages to be measured can swing positive (+ ve) and negative (−ve) we will need to select a bipolar module. Figure 9.4 shows how a potential divider network is used.
- **Current input** By using a 4–20 mA current loop module all the problems of voltage level selection are avoided but at the cost of using process transmitter units to convert the measured variable into current signals. If the transducers have not been purchased at the time of control system specification, devices with built-in current output can be purchased at very little extra cost. The advantages are simpler wiring, better noise immunity (particularly over long distances) and avoidance of earth loop problems. The disadvantage is that most current loop systems have a slow response and are often only suitable for signals which change at a rate less than 10 Hz.

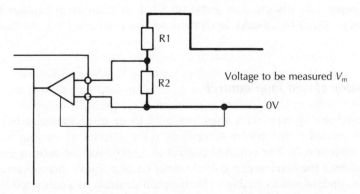

A potential divider network can be used to enable an analog input module to measure a voltage greater than its working range. The maximum voltage that can be measured will be

$V_m = V \times \dfrac{R1 + R2}{R2}$, where V is the maximum input capability of the module.

Fig. 9.4 **Input potential divider network.**

- **Conversion speed**   There are two basic type of A/D converter. The first will perform a conversion every 20 ms (the period of the a.c. mains voltage), which gives us a good clean reading free from worries of line frequency interference. The second will convert in 2–20 µs, giving the possibility of measuring transient data. When a module has more than one input to measure it will do a conversion of each in turn, reducing the data rate for each individual channel. The choice comes down to the number of readings per second we need to capture. This will be the sum of the conversion time and the PLC code required to read and store the data. At very high rates the PLC may only have time to act as a data logger, storing the data as it is read, and analysing it some time after the event to report on or display it. When a very high rate is required we may need to fit an entire module and configure it to read one channel only. If this is not fast enough an external digital voltmeter may be used to take readings in real time at up to 2000 per second and then load them into the PLC using a serial link.

Outputs

The conversion speed of an analog output is generally $< 100\,\mu s$ and rarely a problem. Once the resolution of the module is selected we have only to consider the following points:

- **Voltage level**   Most modules provide ±10 V outputs which can be scaled with a potential divider to the required level. Unipolar devices of 0–10 V are sometimes available, which effectively doubles the resolution when using the same D/A converter.
- **Load resistance**   Voltage output modules are not designed to supply more than a few milliamperes. Typically the minimum load resistance is 300 Ω.
- **Current output**   As discussed in section 6.5.1, it is often an advantage to use a current loop output (4–10 mA). Such modules are available for most PLCs.

9.2.3   *Analog closed loop control*

In many cases analog inputs are used not only to monitor variables but also as feedback to control a process by controlling relay outputs or varying an analog output (see section 6.7). The required control accuracy must be defined in the specification, as must the maximum required rate of change. From this we can calculate the minimum update rate to enable us to maintain control. As an example, consider a storage hopper: its level must be maintained at ±10 mm and it can be filled at a maximum rate of 5 mm/s. A good rule of thumb, to maintain control, is to ensure that the maximum PLC scan time and the input read speed are to be no more than 0.1 of the time required to pass through the ±10 mm control limit. (Time between limits = 20 mm ÷5 mm/s = 4 s. Therefore, scan speed required is 4/10 = 0.4 s.)

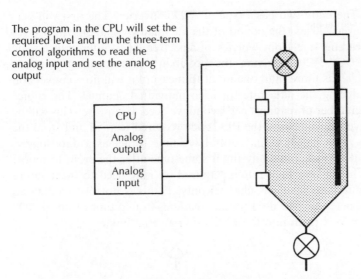

The program in the CPU will set the required level and run the three-term control algorithms to read the analog input and set the analog output

Fig. 9.5 Three-term control using PLC software.

The scan speed must be estimated from the program size, taking into account the PID module (see section 6.7), the execution speed of which will be published by the supplier. If more than one PID loop is to be run at the same time, this execution speed must be multiplied by the number of loops to be run. Figure 9.5 shows the configuration for a PLC implementation of a three-term controller.

If the combined execution speed is unacceptable, a separate PID control module will be required which will handle all the measurements and corrections without increasing the scan speed of the host PLC. Such modules (Figure 9.6) simply require the main program to provide set points (targets) and PID loop constants.

Some PLCs do not support such modules, and in such a case a remote module can be used. These are produced by all the process control instrumentation companies. They require the constants to set up manually and, in use, require an enable signal (relay) and an analog voltage to provide the required set point. Figure 9.7 shows a typical setup.

9.2.4 *Counters, encoders and positioning*

In order to select the correct hardware we need to consider (a) the speed, (b) the total number of pulses to be counted, and (c) the positioning accuracy.

- We must ensure that the electrical characteristics of the PLC and the encoder are matched. The two most common electrical specifications for encoders are 5 and 24 V levels. The 24 V system has the advantage of simplicity but has a relatively

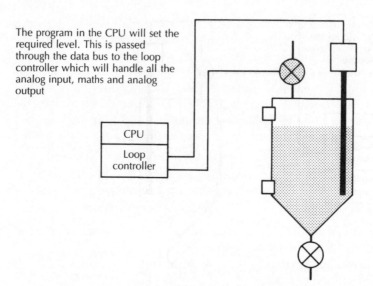

The program in the CPU will set the required level. This is passed through the data bus to the loop controller which will handle all the analog input, maths and analog output

Fig. 9.6 **Three-term control using separate PLC module.**

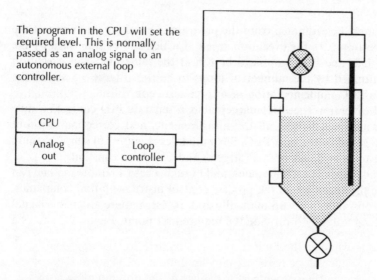

The program in the CPU will set the required level. This is normally passed as an analog signal to an autonomous external loop controller.

Fig. 9.7 **Three-term control using external control module.**

low speed capability (max 25 kHz). The 5 V system, while being somewhat more complex, is capable of 500 kHz switching speeds. The key is to select the PLC module and encoder at the same time to match your application and the voltage levels you already have available.

■ We need to ensure that the system is capable of responding quickly enough not only to count the pulses, but to respond to our target being reached. If, for example, the count information is being used to monitor progress as part of a positioning system, the final stopping accuracy is not only affected by the number of encoder counts per millimetre but by how quickly the system can stop the movement once the correct position is obtained. If we have to rely on the PLC program to scan the count and stop the movement, the error may be significant when the scan time is long and the movement is fast. For example, if the scan time is 100 ms at a traverse speed of 100 mm/s, the uncertainty of seeing and then stopping the movement will cause up to 10 mm (100 mm/s × 0.1 s) positioning error.

There are many possible solutions available to us, each of which offers a trade-off between speed of movement and positioning accuracy. The following list covers a number of solutions with increasing cost, complexity and performance.

1 Use two counter set points, the first to cause the program to change from fast to slow speed, while the second stops on position.
2 The counter may issue an interrupt to the PLC as the count is reached, removing the error due to scan time uncertainty.
3 Use an encoder module with built-in relay outputs to operate the speed change and stop without program intervention.
4 Use a servo control module which will continuously monitor speed and position, adjusting the speed demand to achieve maximum possible speed and positioning accuracy. Such modules normally achieve the performance by using a dedicated processor running a three-term algorithm (see section 6.6.2).
5 If a module is not available for the PLC in use, separate positioning systems are available from motor manufacturers. Generally, they accept commands through a serial link, or are preprogrammed with the required sequence of movements which can be selected and triggered by relay closures.

9.2.5   *Communications*

All PLCs have some sort of built-in communications facility through the programmer port. When we are selecting a PLC where a communications facility is required we need to determine whether the built-in port is adequate for the application, or whether a separate module will be required. Any limitations are likely to be imposed by the facilities the supplier has provided for the use of the programming terminal.

- **Built-in ports** Most PLC suppliers provide the facility to use this port to communicate through a local area network with other PLCs of the same type. In many cases, however, the PLC can only be a 'slave' on the network. A separate PLC or computer must be present which has the role of 'master' in organizing the data transfers from master to slave, slave to master and slave to slave.
- **Add-on ports** These provide the facility for PLC programs to print reports on printers or VDU screens. Section 6.9 covered an example of how this can work. In

addition, some ports allow the much more complex task of setting up a point-to-point communications link with another PLC or computer of any type or manufacture which supports a serial port. Some manufacturers provide software packages to aid this task: in general, however, the user will have to design and implement all the message-passing and error-checking requirements for the safe passing of data.

■ **High-speed LANs**   The operating speed of the data links and LANs described above is relatively slow. The actual port may be capable of around 1000 characters per second but the actual usable data transfer rate is often an order of magnitude lower. If such a LAN has 32 PLCs on it, the delay before any particular node can pass a message can be up to 2 or 3 seconds. When we need to transfer large amounts of data, or when delays before transfers take place cannot be accepted, higher speed LANs are required. Most suppliers provide modules to facilitate such transfers for the larger PLCs. Each has its own system, with its own name but many are based on the Ethernet standard which passes messages over coaxial cable or twisted pairs at a rate of 1 million characters per second.

9.3   Choosing the correct processor

9.3.1   *Capacity*

We now know the types of I/O modules we require and how many of each type. The PLC processor will, however, only be capable of working with a limited number of each type of module or port. As an example, we will consider the Siemens 100U system shown in Figure 9.8. This is a small expandable system which avoids the need for a conventional rack by using a pluggable data bus connecting individual modules fixed to a mounting rail.

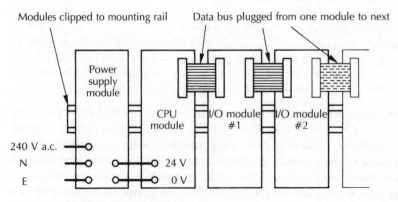

Fig. 9.8   **Siemens 100U bus system.**

This system, using the most powerful processor, can address up to a total of 32 modules. These can consist of simple I/O modules or more complex modules: analog, counters, communications, etc. Only 8 of these, however, can be of the complex type.

9.3.2   *Functionality*

All PLCs have the capability to be programmed to perform the logical operations required for the control of plant and sequences, as discussed in the first five chapters of this book. Such operations require instructions which read inputs, set outputs, perform logical ANDs and ORs, etc. These are normally displayed on the screen and paper as the familiar ladder diagram. To perform other operations further instructions are required and for any particular application it is essential to ensure that the PLC selected can handle the required operations. The following list covers some of the more common instructions and states when they may be used.

- **Data word/byte read/write**   Data transfers and manipulation reading/setting analog values.
- **Multiply; divide; add; subtract**   Arithmetic operations, often on analog or time data. Normally 16-bit integer maths (i.e. the maximum numbers that can be handled simply are in the range -32768 to 32767).
- **Data base operations**   Some PLCs offer instructions which allow searches, sorting and extraction of data from a data base set up in the PLC memory. For example, a data base may be set up to store test specifications. These can then be found and used through a search on the specification number.
- **Floating point calculations**   Some high specification PLCs contain a set of floating point mathematical operators which may include not only $+$, $-$, $\times$ and $\div$ but trigonometrical functions, roots, statistical functions, etc. Floating point notation allows very large (or small) numbers to be manipulated (typically 10^{-38} to 10^{38}). Floating point operations are rarely required.
- **Subroutine calls**   Good system design, as already discussed, requires us to consider the program as a series of interconnected modules. A subroutine facility enables us to implement such a design directly.
- **Data conversion**   Directly converts binary data to binary-coded decimal (BCD) and vice versa. (BCD is a method of expressing data used by many I/O devices.)

9.3.3   *Program speed and size*

As discussed in section 6.6, the maximum speed that any input is read or any output is switched is equal to the program scan time. The required operating speed for all the I/O must be determined, with a PLC selected to match. This is not necessarily an easy task. The scan speed is dependent upon the speed of operation of the PLC and the length of the program. In addition, different PLC operations take different lengths of time, data manipulation taking longer than logical operations. This requires the estimation of the program size and the proportion of slow instructions. Once the program size and type are determined, reference can be made to published scan speeds and a PLC with an appropriate memory capacity and speed can be selected. The scan speed is normally expressed in terms of ms/K (milliseconds to run a 1024-bit instruction) for a stated mix of simple and complex instructions. If we have an unusual mix of instructions we may need to look in more detail at the timings of the instructions we are using to calculate the scan speed more exactly.

9.3.4   *Estimation of program size*

The size of program is dependent upon the complexity of the control problem and the skill and style of the programmer. However, the best program to solve any given problem is rarely (if ever) the smallest. It is more important (and quicker and cheaper) to write a simple easily understood and maintainable program than to strive after an efficient or so-called 'tight' solution.

There is a relationship between program size and the number of inputs and outputs in the system. It is not, however, a direct relationship, the memory usage is increased by increasing logical complexity and any extra functions provided. The following estimation data has been empirically extracted from a number of differing PLC programs and will give a guide. It must be pointed out, however, that the results are only a rough estimate.

A worked example of this procedure can be found in the Appendix, which can be read in conjunction with this explanation.

1 Count up the number of inputs and outputs the system needs. If an operator interface panel is used, each message displayed and key input are also counted as if they were inputs.

2 We now need to scale this by a factor which estimates the complexity of the program. This is achieved by multiplying the count by an estimate of how many times on average each I/O line is to be used for a different purpose in the program. This will produce a corrected count of the number of I/O points used in the program, which we will call COUNT.

   ▪ If a warning lamp were to be flashed at differing rates to indicate five different status conditions, it would require the same amount of code as five separate single-purpose outputs.

   ▪ A limit switch may be used in one section of code to indicate that a movement is finished. It may also, however, be used elsewhere as an interlock to allow or inhibit a separate movement. The effect is to use the same amount of code as two separate inputs.

3 Calculate the base size in bytes by multiplying the COUNT by 15 to obtain the SIZE. This is the size the program would be if we were simply writing a minimum simple program to control the sequence of a mechanism. However life is never

that simple, there are always some extras that need to be added to make the system usable.

4 Each complex I/O point (ANALOG) requires considerably more memory than a simple I/O line. This code is estimated by multiplying the number of points (analog I/O, encoders, etc.) by a byte count of 200.

5 Add the number of bytes calculated so far to determine the program size for the basic program (SIZE).

6 We now need to determine the extra code required for each of the following features which will be required in the program. This is achieved by multiplying the basic SIZE by the following constants. These 'extras' are then added to the base size to obtain the total size of the main program.

Manual sequence as well as Auto (MAN) . . . . ×0.25
Auto restart sequence (REST). . . . . ×0.25
Diagnostic error messages (DIAG) . . . . ×0.5
Test rig? (TEST) . . . . ×1.0
Calibrate sequence (CAL). . . . . ×0.5

Having determined the size of the program to support all the I/O points we now need to estimate our code length to support the more complex features.

7 Determine the quantity of memory required for data storage simply by counting your variables and multiplying by 2 (each variable is likely to require two bytes). The total PLC memory required to hold the data and manipulate it is then estimated by multiplying by a constant. If the PLC has no specific file-handling functions, multiply the data size by 3. If such functions are available, multiply by 2.

8 If an operator display is to be used add 20 bytes for each message *plus* 30 bytes for every variable field used.

9 If a data entry terminal is in use add 300 bytes *plus* 30 bytes for each variable input.

10 If data log or other variable reports are generated add 20 bytes for each fixed line of text with 30 bytes added for each imbedded variable field.

11 If a computer data link has to be written add 400 (COMMS) bytes.

SIZE . . . . ._____
SIZE × MAN . . . . ._____ +
SIZE × REST . . . . ._____ +
SIZE × DIAG . . . . ._____ +
SIZE × TEST . . . . ._____ +
SIZE × CAL . . . . ._____ +
DBASE . . . . ._____ +
MESSAGE DISPLAY . . . . ._____ +
DATA ENTRY . . . . ._____ +
REPORT . . . . ._____ +
COMMS . . . . ._____ +
   Memory required   =

Finally, add the published size of any proprietary software you are using. Once our estimate is complete good practice suggests that we leave at least 20 per cent of the PLC capacity unused to allow for future modifications.

9.4   Selecting a supplier

The choice of supplier for a PLC for a special application is seldom simple. Not only is it essential for the equipment selected to be capable of meeting the technical requirements, but there are a number of other issues that must be addressed. Price is the most commonly stated reason for making a choice, but the true price of a PLC to meet the requirements of a particular application is often much the same over a wide range of supplier equipment.

9.4.1   Functionality

We will have determined the functions and speed required by the PLC and the software during the analysis (decomposition) of the application. We now have to match the application requirements with the features of each of the contending suppliers' equipment to identify the one that best meets our requirements.

9.4.2   Support

Manufacturers of industrial equipment have traditionally given a good level of support to customers. Before any purchase is made the following points should be confirmed with any manufacturer:

- Training
- Technical support (on site and over the phone)
- Application support to configure and design a system
- Rapid exchange/repair of failed equipment
- Guaranteed support for any products for at least 10 years from purchase.

9.4.3   Customer acceptability

In the majority of cases the engineers who design and program a control system are not the end-users who will use it on a day-to-day basis once it is commissioned. In many cases such end-users express a preference for a particular PLC, not because it is technically superior, but because they are already using that type of equipment and have the skills available to keep it running. Such a position is quite valid; while all PLCs are very similar at a conceptual level, the differences at detail level, particularly when looking for faults, are sufficient to require maintenance engineers to be trained and then kept familiar with a system through regular contact.

Some customers go one step further and specify not only the hardware but the actual program structure. This greatly eases the maintenance task of an engineer looking at the program, even for the first time.

The major danger in situations where the hardware is specified by the customer is that the engineer will simply assume, and not determine, that the hardware meets the technical requirements. We must always make an estimate of the requirements before a job is accepted or a quote prepared. If the PLC is not capable of meeting the technical requirements, the job will fail.

9.4.4   *User knowledge*

Learning to use a new PLC takes time. The greater the experience of the programmer or system designer the better, but we must expect to take at least 1–2 weeks to become familiar with the more complex instructions and the programming terminal and software. In addition, the change to a new PLC will require the rewriting of any standard modules that we have used in the past. It may even require a total change to the overall program design philosophy if certain key instructions or methods are not supported. Programs and techniques that use conventional logic functions will move without any problems to most PLCs. Problems occur when the more complex instructions and methods have been used. These instructions are developed by a manufacturer to meet the needs of its hardware design and are inapplicable on different hardware. In Chapter 10 programming methods other than simple ladder or statement list are considered. If such a method has been used the program design and any standard modules will match the requirements of the method. A different manufacturer may not support the method, or implementation may be different. A case in point is 'Grafset' programming (section 10.3.2). Many manufacturers support an implementation (all with different names). They all look very similar at the overview level, but in use they are somewhat different because of the difference in the realization. These differences even affect such major functions as moving between auto and manual control. Such changes must be identified and an estimate made of how much it will cost to use the different implementation.

9.4.5   *Cost*

The cost of a control system is not simply the purchase price of the PLC and other hardware, it is made up of all the following items:

- Purchase price of PLC
- Programming cost
- Documentation cost

- Development cost
- Installation cost
- Maintenance cost

Start-up costs for a new PLC:

- Purchase price of programming equipment and software
- Training (programmer + maintenance) cost.

These start-up costs can be in the order of £500–£8000 according to the complexity of the equipment purchased and courses attended.

Most suppliers have a number of ranges of PLCs. The cost/performance of this hardware often increases in stages as shown in Figure 9.9. The features available at each step vary from supplier to supplier, giving an advantage for any one application.

We should always take care not to select a PLC that will be close to its maximum potential. Any mis-estimation could cause us to run out of memory or speed. At this point in a project all the choices we could make will be expensive. We also may be forced to make some expensive decisions, perhaps switching to a different PLC, reducing the scope of the project or using a second PLC to perform some of the operations required. Choosing hardware with a similar specification from the next range will cost more but there is an easy upgrade path to enhance the performance, i.e. extra memory, faster processors, multi-processors, etc.

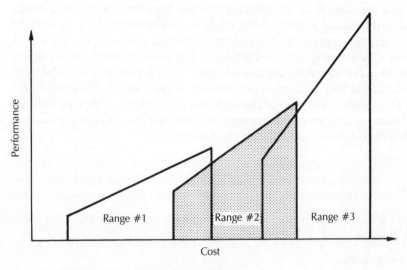

Fig. 9.9 **Cost/performance for PLC ranges.**

9.5   Selection checklist

A very useful *aide-mémoire* during the estimation process is a simple checklist, as shown below.

Job name __*Motor test rig*__   Engineer __*W. Jeffcoat*__   Date __*30/6/92*__

Digital inputs.	D.C. voltage __24__	No. __13__	voltage _____	No. _____
Digital inputs.	A.C. voltage _____	No. _____	voltage _____	No. _____
Digital outputs.	D.C. voltage __24__	No. __7__	voltage _____	No. _____
Digital outputs.	A.C. voltage _____	No. _____	voltage _____	No. _____

Analog inputs.   No. __4__ max. voltage __10__ Accuracy (bits) __12__ Uni-or bipolar __U__

Analog outputs.   No. __1__ max. voltage __10__ Accuracy (bits) __12__ Uni-or bipolar __U__

Counter inputs.   No. ____ voltage ____ Speed (counts/s) _____

Encoder inputs   No. __1__ voltage __5__ Speed (counts/s) __100 kHz__

Servo control   No. __1__ encoder rate Voltage out __10__ positioning accuracy __±1 mm__

3-term control   No. ____ voltage ____ Voltage out ____ min. update rate ____ (s)

Other 'special' modules _____

'BASIC' module   Mem. size ____ comms port type _____

Communications module   Mem. size __4K__ comms port type __RS232__

Advanced program requirements _____ Data handling __*Data base*__

                                          Maths functions __*Analog scaling*__

                                          Floating point _____

Estimated program size _____

Targeted program scan rate _____ __50 ms__

Preferred suppliers __*Siemens*__   __A.B.__

Any special methods required (Grafset; EDDI; etc.) _____

9.6   Summary

The range of PLC suppliers is vast and many offer a number of alternative product ranges with any number of modules, boasting special features. Our choice must meet the application requirements, provide extra capacity for future development and provide a cost-effective solution.

To estimate requirements we have to define the system carefully, and break it down into its main elements. Then we can estimate the number and type of inputs and outputs required.

The size of the program is dependent on the complexity of the control problem and the skill and style of the programmer. Its size can be estimated using a defined methodology.

The final choice of supplier for our PLC is not simple, but will depend upon functionality, support available, customer preferences, user knowledge and price.

CHAPTER TEN

Advanced PLC programming techniques

QUESTIONS ANSWERED

- What programming techniques can we use to get more out of a PLC?
- How can we apply them?
- How can we write understandable, well structured programs?

In the initial section of this book we have covered the traditional operations that give PLC programs their distinctive feel. This chapter is intended for all who have mastered the previous chapters, and need to design a program to control a large complex machine, test rig, process plant, etc.

10.1   Code techniques

10.1.1   *Program order concepts*

A ladder logic program is a software simulation of the operation of control relays, contacts and coils. The most fundamental difference between the two is one of time. In a conventional control panel consisting of relays, as soon as a coil is energized, the contacts will close and energize any other coil in the circuit. The only delay between any two events is the time for current to flow down a wire (speed of light), the electromechanical delay for magnetic flux to build in a coil, and the mechanical delay as the contacts move from one position to another, i.e. if a relay is energized, all its contacts will close together (almost). Any coils connected to these contacts will then all be energized together. In a PLC, however, this is not the case. Only one element of an entire program can be active at any instant of time because the processor can only operate on one instruction of the program at a time. Owing to the high speed of the processor the sequential nature of the process is not normally apparent or important; there are, however, effects which can cause problems or be used to advantage, as shown below.

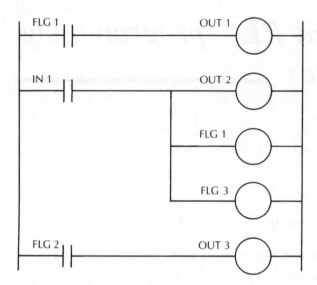

Fig. 10.1   **Example showing effects of scan order.**

Figure 10. 1 shows that, in a physical relay panel, the closure of panel IN 1 would cause OUT 2 to close and its auxiliary contacts FLG 1 and 2 would cause OUT 1 and OUT 2 to be closed simultaneously. In a PLC, however, because the rung with OUT 1 used as a coil is scanned before FLG 1 can be set, the output OUT 1 can only be set one program scan later than OUT 3. An observer would be unable to separate the two events, but if a further line of the program was scanning for both outputs to be ON it would get its result one scan late.

In the following example (Figure 10.2) this effect is used to good advantage to generate a signal that exists for one PLC scan only.

Input IN 1 will set FLG 1 causing OUT 3 to be set through the normally closed contact FLG 2. The final rung will then set FLG 2. At the end of the program scan flags 1, 2 and OUT 3 will all be on. During the next scan, rung 2 will find that FLG 2 is now set, its normally closed contact will be open, causing OUT 3 to be cleared. This is a 'one-shot' or 'first-time flag', frequently used to trigger an event, start a counter or set up data just once when an input changes state.

10.1.2   *Latching outputs*

In all the ladder rungs shown in previous examples the 'coil' will always follow the state of the rung; i.e. if the rung is true the coil is energized, but if the rung is false the coil is cleared. With latched outputs (indicated by a letter S or R in the coil symbol) the coil state is only changed if the rung is true. A set output instruction will SET the output when the rung is true. When the rung goes false the coil will remain on.

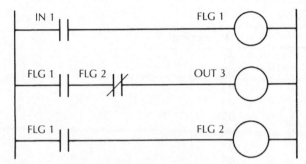

Fig. 10.2 **One-shot generation.**

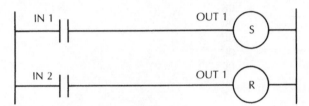

Fig. 10.3 **Latching outputs.**

The fragment shown in Figure 10.3 will set OUT 1 when IN 1 is made. It will then remain (or latch) on until IN 2 is made. This technique 'breaks' one of the rules stated earlier, as we now have a coil existing in more than a single ladder rung. It can be appreciated that this technique needs to be used with great care to avoid very confusing results. The approach can, however, be seen to be both essential and safe when used within program 'modules', as described in section 10.2. In these circumstances we force the PLC to only 'scan' the rungs required to control the operations currently in progress.

10.1.3 *Maths*

Binary numbers

In any PLC information is stored and worked on one word at a time. Each word physically consists of a series of 'bits' (normally 16), each of which will be either ON or OFF. Table 10.1 shows both the relative value of each bit and how any decimal number may be represented as a binary number. For convenience in writing and display they are often expressed as hexadecimal numbers.

Converting a binary number to hexadecimal (HEX), or vice versa, simply requires each digit in turn to be converted from the table above. Conversion from binary to decimal and back, however, requires some calculations to be carried out. Figure 10.4

Table 10.1   **8-bit binary table**

Decimal	Binary	Hexadecimal
1	0000 0001	01
2	0000 0010	02
3	0000 0011	03
4	0000 0100	04
5	0000 0101	05
6	0000 0110	06
7	0000 0111	07
8	0000 1000	08
9	0000 1001	09
10	0000 1010	0A
11	0000 1011	0B
12	0000 1100	0C
13	0000 1101	0D
14	0000 1110	0E
15	0000 1111	0F
16	0001 0000	10
17	0001 0001	11
⋮		
31	0001 1111	1F
32	0010 0000	20
⋮		
255	1111 1111	FF

Each digit in a binary number represents a power of 2. If we consider the following 8-bit binary number 1 1 0 0 0 1 0 1, it can be expanded thus

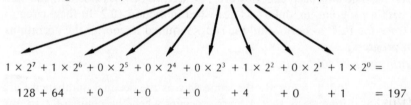

$$1 \times 2^7 + 1 \times 2^6 + 0 \times 2^5 + 0 \times 2^4 + 0 \times 2^3 + 1 \times 2^2 + 0 \times 2^1 + 1 \times 2^0 =$$

$$128 + 64 \quad + 0 \quad + 0 \quad + 0 \quad + 4 \quad + 0 \quad + 1 \quad = 197$$

Fig. 10.4 **Construction of a binary number.**

shows how a binary number is constructed. All we need to know is the table of powers of 2, which is given for 0 to 15 (16 bit numbers) in Table 10.2.

To convert decimal numbers to binary we need to use the reverse process. We subtract powers of 2 (starting with the largest) in turn from the number to be converted. If the power is less than the number to be converted, we insert a 1 into the appropriate position of the answer while the remainder is carried through to the next stage. If the power is less than the number, we insert a 0 in the answer and the unchanged number is carried through to the next stage. Consider the conversion

Table 10.2 **Powers of 2**

Power of 2	Decimal number	Power of 2	Decimal number
0	1	8	256
1	2	9	512
2	4	10	1024
3	8	11	2048
4	16	12	4096
5	32	13	8192
6	64	14	16384
7	128	15	32768

Table 10.3 **Decimal to binary conversion**

Powers	Calculation	Stage result	Binary answer
143 − 2 =	143 − 128	1 remainder 15	1 − − − − − − −
15 − 2 =	15 − 64	0 remainder 15	1 0 − − − − − −
15 − 2 =	15 − 32	0 remainder 15	1 0 0 − − − − −
15 − 2 =	15 − 16	0 remainder 15	1 0 0 0 − − − −
15 − 2 =	15 − 8	1 remainder 7	1 0 0 0 1 − − −
7 − 2 =	7 − 4	1 remainder 3	1 0 0 0 1 1 − −
3 − 2 =	3 − 2	1 remainder 1	1 0 0 0 1 1 1 −
1 − 2 =	1 − 1	1 remainder 0	1 0 0 0 1 1 1 1

of 143 to an 8-bit binary answer (Table 10.3). It can be seen how in each stage the remainder is carried down to the next while the answer of the subtraction is used to form the binary number.

Addition and subtraction

Binary numbers are added in exactly the same way as the decimal numbers we are used to. The only difference is that a 'carry' occurs when the count of 1 in any position occurs rather than a count of 9. Figure 10.5 demonstrates the principle. Subtraction is normally carried out by adding a negated number, i.e.

$5 - 3$ is the same as $5 + (-3)$

Numbers are negated by generating their 2's complement. This is a two-stage process: (1) the complement is formed by 'flipping' each of the bits (1s → 0s and 0s → 1s); (2) the result is then incremented by 1, i.e.

$\quad$ 25 dec = 0 1 1 0 0 1 $\quad$ binary; this is complemented to form
$\quad\quad\quad\quad\quad\quad$ 1 0 0 1 1 0 $\quad$ which is in turn incremented to give
$\quad$ −25 dec = 1 0 0 1 1 1 $\quad$ binary

Fig. 10.5 **Binary addition.**

Table 10.4 **Range of 16-bit binary numbers**

Binary table	
7FFF	32767
7FFE	32766
.	
.	
0002	2
0001	1
0000	0
FFFF	−1
FFFE	−2
FFFD	−3
.	
.	
7FFF	−32767
8000	−32768

Negative numbers can always be identified by the most significant bit being set to 1, i.e.

	4 bit	8 bit	16 bit
−1 dec =	1111	1111 1111	1111 1111 1111 1111

Table 10.4 shows the range of numbers available in a 16-bit PLC. The convention of the sign bit is in some respects a little artificial and does cause some problems. Adding 1 to 32767 will not give the expected result 32768 but −32768; by adding one more we get the answer −32767, etc. It is not uncommon to ignore the effect of the sign bit and use the full range of the 16 bits as a positive number (0–65535 [2^{16}]). In most cases this will cause no problems unless use is made of multiply, division or comparison routines which are expecting signed numbers as the input. In such cases the results returned will not always be as expected.

Where results are required that have greater accuracy than the PLC word will support, we are forced to use two or even three words of memory to store and manipulate our data. To facilitate such operations the PLC is equipped with a carry or overflow flag. This is set whenever the result of an addition causes an overflow of the result, i.e. in a 16-bit PLC 65535 (FFFF hex) + 1 = 0 + carry. The carry is then added to the next most significant word of the data before any addition is performed on it.

Multiplication and division

If multiplication and division instructions are not available in the PLC to be used for a particular application, the required functions can be written using the add and subtract instructions. Such a code is, however, very expensive to write and easy to get wrong. It is cheaper to purchase a PLC with the required functions or purchase the software required from your supplier.

Representation of fractions

Binary number representation only allows integers (whole numbers) to be represented (i.e. fractions cannot be expressed). A common requirement is to work to one or two places of decimals. For example, we may wish to read, scale and subsequently print a voltage reading to two decimal places. The simple way to handle this situation is to scale the input and any subsequent calculations in units of 1/100 volt, so a reading of 11.45 volts would appear in the PLC as 1145 (decimal). This also has the advantage of making any intermediate calculations easy to follow as we can easily relate the numbers in PLC registers to the measured 'real world' voltage.

10.1.4   *Interrupt scans*

As discussed in section 6.6, hardware generated interrupts are used to ensure rapid response to an event. In a simple example we may wish to clear an output *immediately* a switch is made, but only if two other events (remembered in the PLC by flag bits FLG 1 and FLG 2) have occurred.

    If the rung were in a normal program it would execute once every scan of the PLC. If the scan speed was 100 ms, there would be a delay of between 100 and 200 ms (see section 6.6), from the setting of the input IN 1 to the setting of the output OUT 3 by the PLC. If an interrupting input line has been used this fragment of code would be given a specified name or placed at a known address within the PLC. The input IN 1 changing state would cause the PLC to stop its program scan and execute this fragment once only. As described in section 6.6, the inputs the program reads are in reality a record of their state before the start of the program scan. When responding to an interrupt we are provided with the facility to read the actual state of the input and to immediately set or clear any outputs. Interrupt response facilities are typically provided for the following input types:

- Digital inputs
- Counters
- Encoders position modules
- Communication modules.

Once an interrupt program is in progress the processor cannot respond to another until the first has been fully serviced. To reduce the subsequent danger of extending the response times, interrupt handling fragments must be kept as short as possible. They should carry out only the essential task required and set a flag to trigger the main program to carry out any housekeeping.

There are instances where a few rungs of the main program must be protected from an interrupt occurring within them. An example of this could be where a series of data elements which are updated by an interrupt source, are to be sorted. If an interrupt occurred part of the way through the sort, the results would be totally invalid. The disabling of interrupts must be kept to an absolute minimum or the advantages of rapid response are lost.

10.2   Modular programming techniques

One of the features of ladder programming is that it shows at a glance the status of a screen full of inputs, outputs and flag bits and how they affect each other. It is, however, very poor at showing the overall structure, i.e. how all the interactions and logical steps that constitute the control logic for a particular piece of equipment fit and interact together. Chapter 7 has described how we should design programs as a series of interacting program blocks. This section will describe a number of techniques that can be used to implement modular programs.

10.2.1   *Master control relays*

A master control relay (MCR) is perhaps the simplest technique we have to enable a ladder program to be written as defined blocks. The fragment shown in Figure 10.6 presents a block that will be scanned while flag bit FLG 1 is set. When the flag is clear the code within the block is skipped by the processor. This technique has the added advantage over a non-block structured program of reducing the time required to scan any given program because (in most PLCs) only the currently required code is scanned. In some PLCs, however, the instruction operates by scanning all the rungs but forcing the entire rung to a FALSE condition.

Such blocks are normally placed together at the end of the program with the controlling logic defining the program structure placed at the front.

When such a block of program is required to receive and process data it must be stored in a defined register or memory at the same time the controlling flag is set. The block can then access the data when it runs and store the result. It is possible to use such data-passing techniques to write a block that may perform a task required in many parts of the main program, such as scaling analog results.

The format of the instructions does of course vary from supplier to supplier, but the concept is the same in all cases.

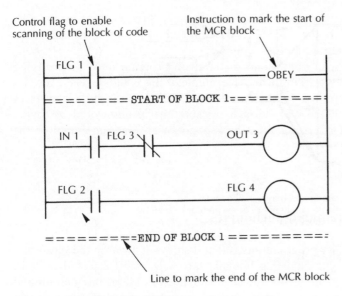

Control flag to enable
scanning of the block of code

Instruction to mark the start of
the MCR block

Fig. 10.6 **Diagram MCR block.**

10.2.2 *Program blocks*

The concept of program blocks has been borrowed from the world of computer programming. It is a much more powerful technique but not universally available. These blocks are subroutines that are 'called' by the main line program. The actual program scan diverts to a called block and then returns to the instruction following the call.

If a block is called four times in the main program, it will be scanned four times in the program cycle. This removes the problem we considered with master control relays concerning ownership of results from calls on such blocks, as the result is available immediately for the program rung following the block call.

It is possible for a block to force a return to the calling program (or block end) before it has run to its end. This is achieved using a conditional block end instruction. This technique can be used to good effect to prevent code being scanned when it is not required. As an example, a program may be designed with four main blocks: test, calibration, data entry and reports. If each is written as a separate block called from a controlling program, the functionality of each is kept separate. Not all of the functions will be active all the time. An enabling flag can be allocated to each function allowing the program block for each to return immediately if it is not required. This has the advantage of speeding up the scan rate (less code to scan).

The calls on program blocks (often incorrectly named 'jumps') can also be conditional on the state of a ladder rung. This serves a similar function to conditional

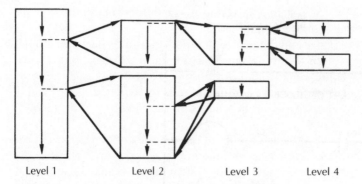

Level 1   Level 2   Level 3   Level 4

This diagram shows how program blocks may be called from one
another. Any block may be called any number of times from any other
block providing the maximum number of levels is not exceeded
(normally 16 levels).

Fig. 10.7 **Program block call levels.**

block ends, but places the decision within the calling program rather than within the
called block.

The calling of program blocks is not limited to just the main program. A called
block may itself call another, and so on (up to the limit specified by the PLC
manufacturer). Figure 10.7 shows the effective program flow as the levels of program
block are 'called'.

10.2.3 Parameterization

When the system decomposition identifies a task needed to be performed many times
but with different data values or I/O lines, a program block may be designed and
written to perform the function of receiving variable data from the main line or
calling program. Such input data may be stored by the program in PLC registers
or memory, or in some PLCs a program block may be used which directly supports
the transfer of data from the calling program into the one being called. Such data is
known as 'parameters'. Many types of data may be passed into such program blocks.
In the Siemens system, for example, they are known as function blocks and any type
of data can be passed (and returned to the calling program).

10.3   Finite-state programming

Many of the systems we control with PLCs are best described as a sequence of
events. They are normally programmed as ladder rungs through convention and
the engineer's knowledge rather than by use of the most appropriate technique.

Finite-state design looks at a system as a series of 'conditions' or 'states' and the events that will cause one to move to another.

Most machine tools, test rigs and data entry or report tasks are effectively a series of actions followed by a test or tests for events to determine if they have been carried out correctly. If we write our code to look for a single event during each state, we build up a simple linear sequence; by the addition of tests for extra events we build in the ability to branch to alternative states dependent on the condition of the plant or operator responses.

It can be seen that, generally, the description of the state describes what action has to be taken, i.e. what valves, flags, messages require to be set (outputs), while the described events to enable mode changes are the system inputs, scanned flags, etc. A practical implementation of the method consists of a program block for each such state. This will have two parts: the condition code, which will perform the required operations for this part of the program, and the termination code, which will look for a predetermined event that will determine when the conditions are ready for the sequence to move on to the next state. If there are more than one set of conditions that can terminate a state, then there will be a separate section of code to look for each.

A finite-state diagram is used to show the program states and the conditions required to move between them. In Figure 10.8 a circle and description are used to represent each state, while arrows show the flow from one state to another when the specified condition is met.

If we consider state 2 of this figure, it describes the conditions of the vehicle moving ahead in a straight line, following a line on the floor. The conditions controlled by it are to set the drive motor on and centre the steering. There are three events to be looked for by the code, which will cause a change of state:

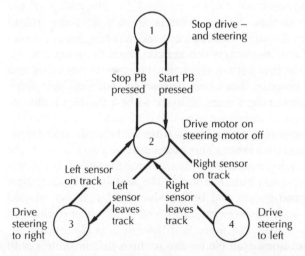

Fig. 10.8 **Finite-state diagram for a simple automatic guided vehicle.**

1 The stop push button being pressed – state 1 is made active.
2 Left sensor determines we need to turn to the right – state 3 is made active.
3 Right sensor determines we need to turn to the left – state 4 is made active.

In such a program only one block of code will be active at any time making it very easy to write and test the code as there are no interdependencies. Finite-state programming has the effect of latching all outputs ON or OFF dependent on the operation of previous blocks. This must be taken into account as each block is written. Where more than one sequence of events are required to take place simultaneously, a separate finite-state program is made active for each.

The management of the blocks may seem difficult but it is in fact very easy utilizing the conditional block structured instructions described previously. A flag can be allocated to each state and set when the code is required to run, while the flag for the previous state is cleared, any code that must always run (stops, guards, etc.) being handled in a permanently scanned block.

The example presented in Figure 10.9 shows the finite-state diagram required to define the paint plant example described in Chapter 12, exercise 12.1.

10.3.1 *Sequencers*

Some manufacturers (notably Allen Bradley) have built into the instruction set of the PLC special instructions to make the implementation of non-branching sequences very easy. The following description considers only 16 input and output lines, but the instructions can be used in parallel to monitor and control any number of lines.

The sequencer output instruction (SQO) will control all 16 lines of an output module or flag word to match the requirements of a sequence. A data file is set up containing each of the 'patterns' of outputs required for every step of the sequence. As the sequence progresses the next pattern is moved to the output port. The sequence is progressed through the steps using the matching input instruction (SQI). This has a similar file of patterns which are compared to an input word. When the input word matches the first pattern the internal counter is moved on and with it the output instruction counter. The input instruction will now wait for a match with the second element in the file. Figure 10.10 shows how the data tables are set up.

When used in this way the instruction could only work if the input and output ports were dedicated to the single sequence. This is because at every step all the outputs will be controlled to match the pattern in the sequence file and all the inputs must be at the expected state to allow continuation to the next step. To overcome this, separate filters for input and output can be introduced. This allows selected inputs to be in a 'don't matter' state and selected outputs to be left in an unchanged condition.

If we again consider our paint plant example we can set up a simple sequencer to control it. Figure 10.11 shows how two sets of flag bits in two flag words have been

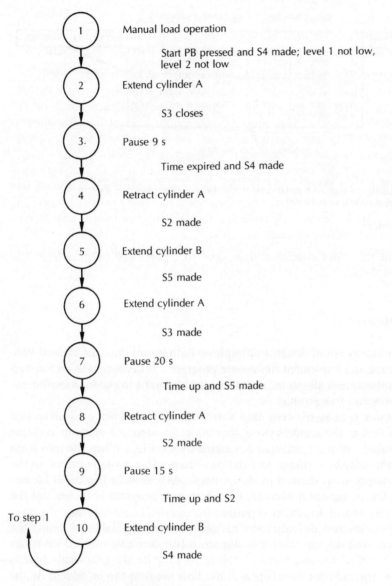

Fig. 10.9 **Paint plant example.**

allocated as input and output to the sequencer. Each input bit is controlled by a physical input or timer contact, while each output bit controls a physical output or timer coil. The pattern of input flags to allow the sequence to continue and the required condition of each output is entered into the condition blocks for each of the 10 steps. A few simple lines of ladder are then required to allow the inputs and

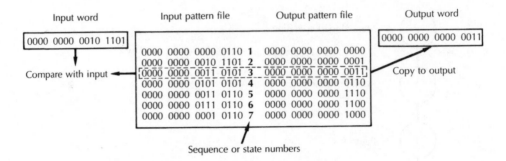

| Input word | Input pattern file | Output pattern file | Output word |

```
Input word                Input pattern file        Output pattern file      Output word

0000 0000 0010 1101                                                        0000 0000 0000 0011
                          0000 0000 0000 0110  1    0000 0000 0000 0000
                          0000 0000 0010 1101  2    0000 0000 0000 0001
Compare with input ◄      0000 0000 0011 0101  3    0000 0000 0000 0011      Copy to output
                          0000 0000 0101 0101  4    0000 0000 0000 0110
                          0000 0000 0011 0110  5    0000 0000 0000 1110
                          0000 0000 0111 0110  6    0000 0000 0000 1100
                          0000 0000 0001 0110  7    0000 0000 0000 1000

                          Sequence or state numbers
```

The input word has matched the state-two input conditions and stepped onto state three, setting the state-three output pattern. The system is now waiting for the state-three input conditions to be met.

Fig. 10.10 **Sequencer.**

timers to control the input condition flags, and to control the outputs from the output condition flags.

10.3.2 *Grafset*

Grafset is a graphically based design and implementation tool that can be used very effectively to define and implement finite-state programs. It takes care of all the step selection housekeeping and allows the user to design and write the code to implement the required states and transition code.

Grafset diagrams consist of steps that form the working part of a state and transitions that define the conditions to terminate the step. At start-up a single step and its transition is made active and scanned by the PLC. When the transition becomes true (normally by virtue of its final coil being set), activity is passed to the next step and transition, as defined in the diagram. As shown in Figure 10.12, not only are simple linear sequences allowed, but alternative program branches and the simultaneous operation of a number of paths.

The diagram shows and describes the various elements available to define the control structure. The actual control is written using conventional ladder rungs which reside in each of the steps and elements. In practice the graphical chart is drawn and each step and transition 'opened' to allow the rungs to be added. As the program runs, the rungs for the active step (or steps) are scanned by the PLC along with its transition code (often a single rung). If the final rung (coil) in the transition is true, its step is closed and the next step–transition partnership as defined by the graphical connections is made active. In some systems the PLC will only scan the code in 'active' steps while in others all the steps are scanned but the rungs forced FALSE in non-active steps.

During program testing and debugging the active steps in the network are highlighted on the programming terminal. This greatly simplifies the fault-finding pro-

Paint plant flag allocation

Input flag (data) word
FLG 1   Start PB
FLG 2   LS2
FLG 3   LS3
FLG 4   LS4
FLG 5   LS5
FLG 6   Level 1
FLG 7   Level 2
FLG 8   Timer 1 output
FLG 9   Timer 2 output
FLG 10 Timer 3 output

Output flag (data) word
FLG 20 cyl. A+ output
FLG 21 cyl. A− output
FLG 22 cyl. B+ output
FLG 23 cyl. B− output
FLG 24 Control timer 1
FLG 25 Control timer 2
FLG 26 Control timer 3

The condition blocks are then set up as below with the input pattern to allow
movement to the next state and the output pattern to be set at each state.
The input condition block is allocated right to left, FLG 1 to FLG 10.
The output condition block is allocated right to left, FLG 20 to FLG 26.

State	Input condition block	Output condition block
1	0000 0000 0110 1011	0000 0000 0000 0110
2	0000 0000 0110 1100	0000 0000 0000 0101
3	0000 0000 1110 1100	0000 0000 0001 0101
4	0000 0000 0110 1010	0000 0000 0000 0110
5	0000 0000 0111 0010	0000 0000 0000 1010
6	0000 0000 0111 0100	0000 0000 0000 1001
7	0000 0001 0111 0100	0000 0000 0010 1001
8	0000 0000 0111 0010	0000 0000 0000 1010
9	0000 0010 0111 0010	0000 0000 0100 1010
10	0000 0000 0110 1010	0000 0000 0000 0110

A set of simple ladder rungs would then be set up to allow input conditions to
control the input flag bits individually and allow their output flag bits to control the
outputs and timers, as indicated below.

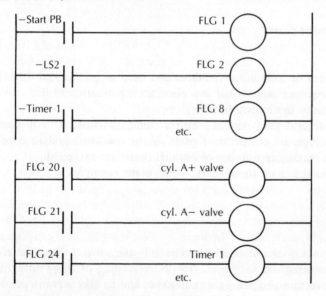

Fig. 10.11 **Paint plant sequencer example.**

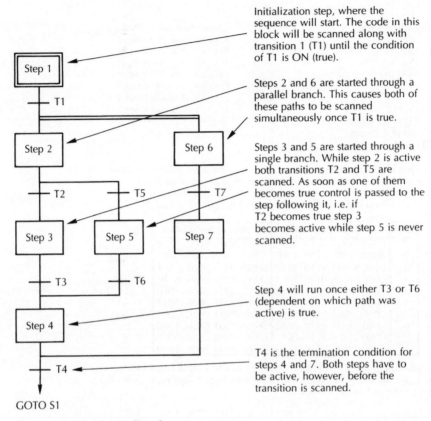

Initialization step, where the sequence will start. The code in this block will be scanned along with transition 1 (T1) until the condition of T1 is ON (true).

Steps 2 and 6 are started through a parallel branch. This causes both of these paths to be scanned simultaneously once T1 is true.

Steps 3 and 5 are started through a single branch. While step 2 is active both transitions T2 and T5 are scanned. As soon as one of them becomes true control is passed to the step following it, i.e. if T2 becomes true step 3 becomes active while step 5 is never scanned.

Step 4 will run once either T3 or T6 (dependent on which path was active) is true.

T4 is the termination condition for steps 4 and 7. Both steps have to be active, however, before the transition is scanned.

Fig. 10.12 **Simple Grafset chart.**

cess as only a few lines of code are ever active and need to be checked for interactions. If a machine sequence stops while in service, an examination of the active step will immediately identify the reason for the stoppage.

In most systems several networks can be run independently. This is useful to further separate the logic of independent parts of the machine, and to cope with situations where the maximum number of parallel paths are exceeded.

Figure 10.13 shows the chart to solve our paint plant example.

10.4   Summary

A PLC ladder program does not function in exactly the same way as the electromechanical relays the program simulates. An understanding of these differences is needed to overcome certain problems that can arise, and to take advantage of their effects.

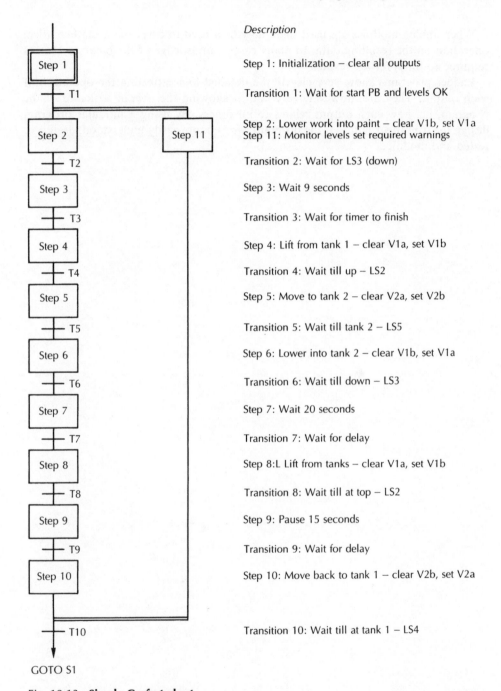

Description

Step 1: Initialization – clear all outputs

Transition 1: Wait for start PB and levels OK

Step 2: Lower work into paint – clear V1b, set V1a
Step 11: Monitor levels set required warnings

Transition 2: Wait for LS3 (down)

Step 3: Wait 9 seconds

Transition 3: Wait for timer to finish

Step 4: Lift from tank 1 – clear V1a, set V1b

Transition 4: Wait till up – LS2

Step 5: Move to tank 2 – clear V2a, set V2b

Transition 5: Wait till tank 2 – LS5

Step 6: Lower into tank 2 – clear V1b, set V1a

Transition 6: Wait till down – LS3

Step 7: Wait 20 seconds

Transition 7: Wait for delay

Step 8:L Lift from tanks – clear V1a, set V1b

Transition 8: Wait till at top – LS2

Step 9: Pause 15 seconds

Transition 9: Wait for delay

Step 10: Move back to tank 1 – clear V2b, set V2a

Transition 10: Wait till at tank 1 – LS4

GOTO S1

Fig. 10.13 **Simple Grafset chart.**

When analog modules are used there may be a need to carry out a mathematical operation on the resulting data. In many cases a knowledge of the binary system is required to interpret the results.

Ladder programs show very clearly the detailed logic affecting the operation of each output. They are, however, very bad at showing the overall structure of the program, i.e. how each output relates to all others. By using a modular program design the overall structure of the program can be more easily understood, designed, tested and modified.

CHAPTER ELEVEN

Fault diagnosis techniques ▬

QUESTIONS ANSWERED

- How can we determine why a program is not working?
- What are fault diagnostic programs?
- How can we program the PLC to diagnose machine faults?
- How can we display the faults identified?
- How can we use our design documentation to diagnose fault conditions?

If you have so far understood the material explained in the previous chapters it must be apparent that a PLC program contains a great deal of personal creative effort. The loss of this program either by accident or deliberate action renders the whole machine/PLC installation useless – a machine without a brain. Unlike hardware-based machines or process plant the loss or corruption of software in a computer-controlled system makes it extremely difficult to restore the system to its production mode. Note that with purely hardware-based controllers it is virtually impossible to 'lose' the control rules designed into the circuits. By substitution of suspect items of hardware, aided by a good diagnostic technique, it is relatively easy to restore the system to an operational state. The bottom line for managers who are responsible for the use and maintenance of PLC-controlled systems is effective software security and support. To achieve this, technically trained staff competent to edit or reprogram the PLCs must be available for the duration of ownership of the equipment. This is even more of a problem if subcontract staff are used in this role, and an even bigger problem exists if the subcontracting company closes down leaving their customers 'in the lurch'.

11.1   PLC diagnostic displays

The vast majority of PLCs have input and output status displays which are provided to help the user to diagnose hardware and software fault conditions. These indicators are of great use in diagnosis, revealing missing (or spurious) signals, lost outputs, etc.

They are, however, of limited use in the diagnosis of machine sequence or software problems unless the user has a full, detailed understanding of the machine operations and the control program. Normally, maintenance staff who are sent to diagnose and rectify a system fault are not present prior to the breakdown occurring and thus are not aware of where in its sequence the machine has stopped. This makes diagnosis much more difficult than the situation in which they were present and able to observe the exact location within the program where the fault occurred.

A number of simple techniques are described below which are aimed at making use of the information within the PLC to simplify the diagnostic activity, which in many large installations poses a serious problem to the management. The vast majority of PLC programs do not completely fill the memory available. This 'spare' capacity can be used to help pinpoint the fault conditions, thus reducing downtime with no further hardware investment.

11.2   Last output set technique

The objective of this technique is to use an additional set of status lamps to indicate the last output set during the sequencing of a machine or process plant. If we know the last operation started we should quickly be able to determine the reason for the fault condition.

The technique is best described by means of an example: A 4-cylinder pneumatic system is being controlled by a PLC, and Figure 11.1 illustrates the configuration of cylinders A, B, C and D. The sequence being $A+ B+ C+ D- C- B- A- D+$, where, as usual, '$+$' means piston extending and '$-$' means piston retracting. The ladder diagram that gives this sequence is shown in Figure 11.2, which can be termed an 'executive' program since it executes commands to the machine being controlled. During normal, fault-free operation, the configuration of piston position steps from the initial state to its final state can be seen to be identical, as illustrated in Figure 11.3. The problem for a technician when confronted with the system after a fault has caused the sequence to halt at some part of its program, is: Where in the sequence has it actually stopped? If the technician is watching the machine operating and, for instance, has observed the sequence $A+ B+ C+$, and the machine then stopping, he or she knows that the next operation that should have occurred is $D-$. Effort could then be concentrated on determining if the operation has not occurred because of a failure of the $D-$ solenoid or whether the conditions to initiate the $D-$ movement have not been achieved. We can add code to effectively 'watch' the sequence for us and indicate how far we have progressed. This is achieved by taking the existing operation (executive) outputs and, after conversion into a short duration pulse by means of a timer (or a single-shot output), latch each output onto an array of additional diagnostic LEDs. These will at all times show the last output set by the program, immediately pointing to the area where the fault exists. This status display is of considerably greater use than the traditional pattern of input/output LEDs

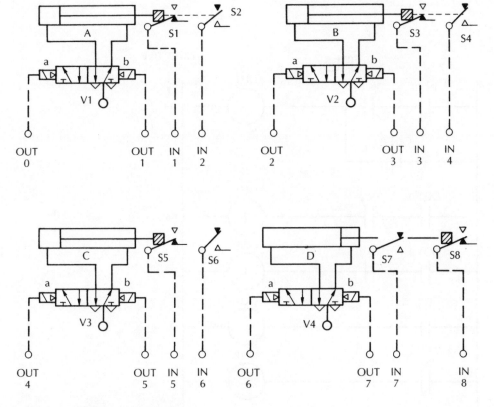

Fig. 11.1 **Configuration of pistons.**

which then need to be interpreted by the technician with the aid of the ladder diagram and logical detection.

11.2.1 *Detailed operation of the technique*

In Figure 11.2 the 'START sequence' switch is operated closing IN 10 (top rung of ladder diagram), OUT 0 is produced and remains ON through the operation of the latch circuit OUT 0 contact. OUT 0 gives rise to cylinder A+ motion, which in turn closes IN 2 giving us the B+ operation. In addition it can be seen from Figure 11.4 that OUT 0 also produces a short-duration pulse at auxiliary output R20 through the action of timer T2. Referring now to Figure 11.5, Output R20 now completes the diagnostic circuit causing output LED10 to be illuminated indicating that the A+ operation should have occurred. If A+ operates (refer back to Figures 11.2 and 11.4) the IN 2 makes, and OUT 2 results. This in turn produces OUT 22 which 'unlatches' OUT 10 (Figure 11.5) and also latches the next diagnostic OUT 12 (= B+).

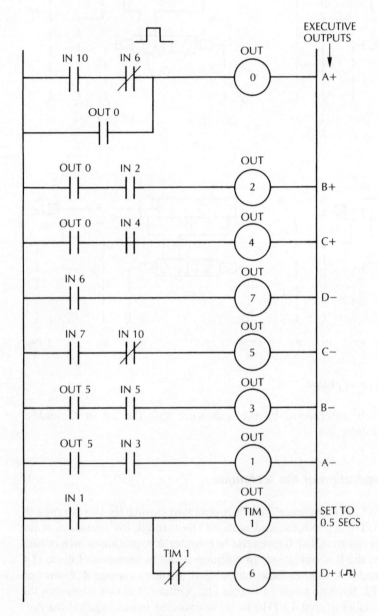

Fig. 11.2 **Executive program.**

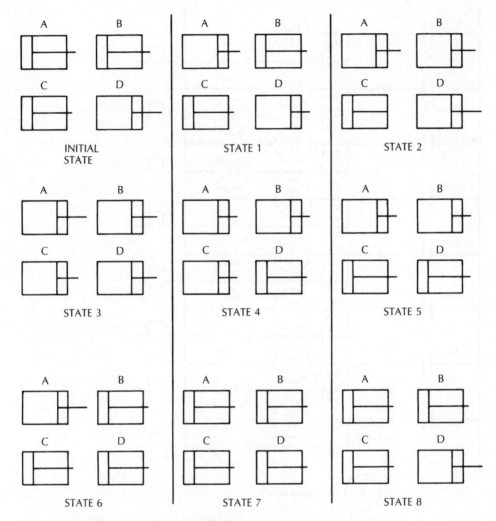

Fig. 11.3 **Configuration of ABCD cylinder.**

Similarly, as each new output is turned on any previous diagnostic output is cleared and a new one is set. If a fault develops then the last successful output will remain ON indicating where the sequence has stopped; e.g. if C+ solenoid became unserviceable prior to the sequence being operated, output LED 14 will illuminate but the sequence will have stopped at the B+ position. Only one diagnostic display will be ON at any one time, and in this example the LED 14 being ON indicates that the C+ operation is not completed.

If the existing cylinder sequence were completely changed, then the order of operation of the diagnostic LEDs would change automatically to follow the changes

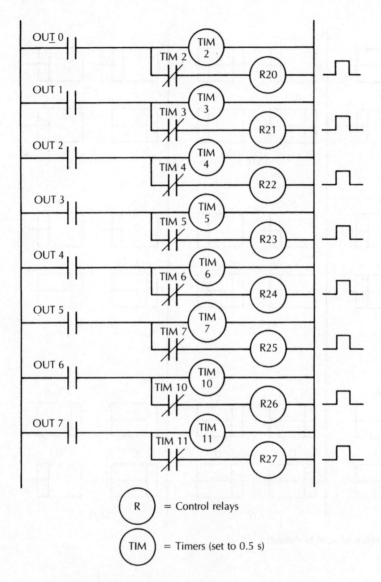

Fig. 11.4 **Diagnostic program – timers.**

in the executive program. As before, the only diagnostic LED being ON would indicate the last successful output set. As can be seen in Figure 11.5, the diagnostic program is independent of the operating sequence of the outputs and thus only the executive program would need to be changed.

There is, however, a limitation inherent with this technique that needs to be identified. The method is designed to 'latch' the final output set on. If more than

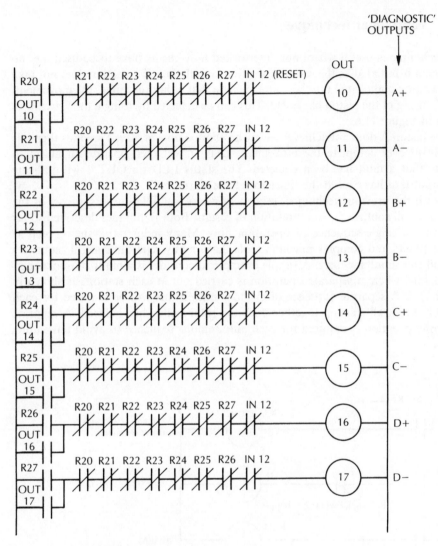

Fig. 11.5 **Diagnostic program – outputs.**

one output is set ON together, or a second is set ON during the pulse time of the first, they will mutually clear each other's diagnostic outputs until the timers finish. At this point one of the diagnostic LEDs will be set ON. Should one of these cylinder operations now fail there is no guarantee the correct LED will be ON. Indeed, should both timers finish in the same PLC scan, no outputs at all will be set!

11.3   Fault timer technique

This technique is useful where non-interlocked movements have to be used, i.e. no movement is initiated as the result of the first completing. If an action does not take place within an allocated time we can conclude that either the valve, the movement, or the sensing of the piston has failed. This can be detected by using a time circuit, as shown in Figure 11.6.

If the piston A does not achieve A+ by an elapsed time of, say, 3 seconds then it is unlikely to ever complete the movement. OUT 7 will then operate an alarm to indicate that a fault has been detected. The status LED on OUT 1 will indicate the output that has caused the timer to set the alarm.

As with the previous technique, sequences where multiple outputs are set at the same time will confuse the interpretation of results. Both techniques described so far work with a single sequence of operation only. Many industrial systems, such as process plants and complex machine tools, have many separate control sequences that will run simultaneously. A simple example to consider is a multi-station assembly machine where a separate operation is carried out at each station, as shown in Figure 11.7. A separate sequence of cylinder movements will be required for each assembly head, all starting together once the rotary table has indexed. A separate diagnostic program is required for each autonomous sequence to avoid misleading results.

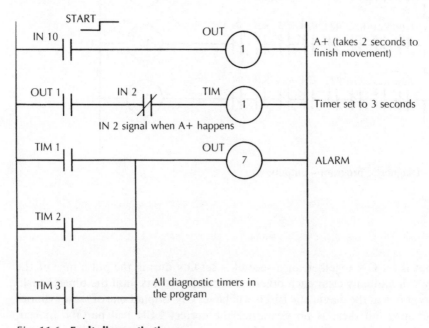

Fig. 11.6 **Fault diagnostic timers.**

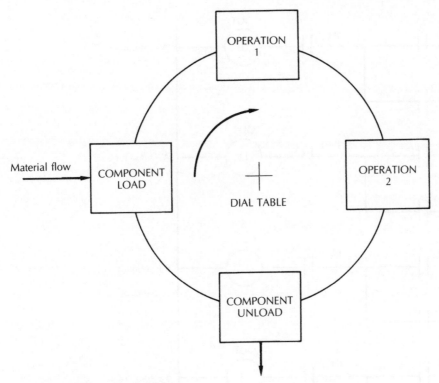

Fig. 11.7 **Multi-station machine.**

11.4   Uncompleted operations technique

As seen in all the examples described so far, the vast majority of operations controlled by PLCs result in some form of input back to the PLC to indicate that the operation is completed. It is simple enough to arrange for a status output to be latched by the start of a movement and then cleared by its completion. A suitable circuit is shown in Figure 11.8 to add such facility to monitor cylinders A and B in the example described earlier in Figure 11.1.

This diagnostic display will show any uncompleted movements under all conditions. If the plant stops, the display will indicate the movement (or movements) at error. If required, an ON delay timer can be triggered by ORing all the status bits, producing a warning output when any diagnostic output stays on beyond the cycle time of the machine.

Like the first technique, if the executive program has to be modified to change the operational sequence there is no need to change the diagnostic program. It will continue to work providing the physical arrangements of cylinders and sensors are not changed. This technique does not suffer from the problems of the first two

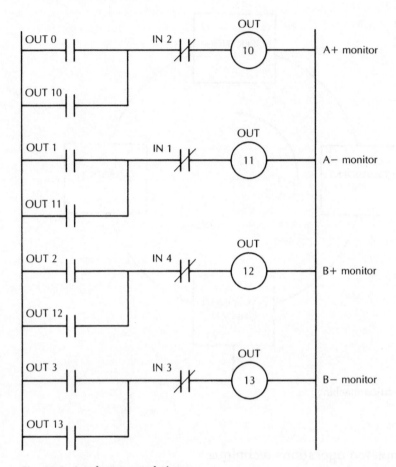

Fig. 11.8 **Latch status technique.**

regarding simultaneous movements, the status display always indicating uncompleted operations.

There are, however, areas of operation where the situation is not so easy. Cylinders, particularly hydraulic cylinders, are not always driven to the end of the stroke. The example in Figure 11.9 shows a cylinder used to control an unload arm. The gripper is moved to a 'wait' condition while an operation is carried out, and then moved again to its full extent to grip the component to be unloaded.

The ladder rung to control the cylinder movement must consider both sets of conditions, as must the diagnostic rung. Figure 11.10 shows the rungs to cope with this situation. If the sequence of operations were to be changed, however, we would have to carefully examine the diagnostic code to determine any modifications required.

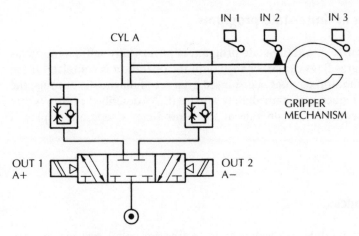

Fig. 11.9 **Gripper arm arrangement.**

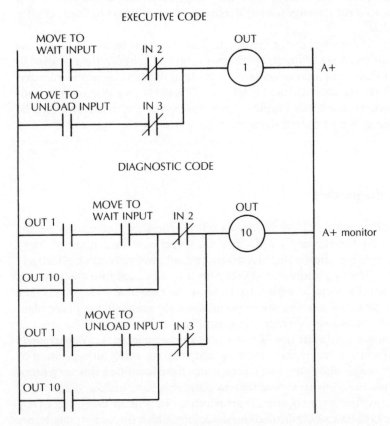

Fig. 11.10 **Control and monitoring of gripper arm.**

11.5   Diagnostics in finite-state programs

As discussed in section 10.3, a finite-state program is intrinsically self-diagnostic as the step counter or status flags cannot change until the operation is complete. If the step control flags or the stop counter is displayed on a set of diagnostic outputs, the current status of the sequence is immediately apparent. As described above, where more than a single sequence can operate at the same time, a separate display is required for each.

11.6   Fault messages

The use of an output module to display faults is fine for small systems but can become impractical on large plants where several separate outputs modules may be required. If an operator interface, as described in section 6.9, is available it can be used to display an error message for any diagnostic control relays or flags set after a specified time delay.

If such a display is not available, control relays or flags can be used to store the diagnostic information. A set of inputs (perhaps a rotary switch) can then be used to 'select' a group of flags onto a set of outputs set aside to display diagnostic information. Figure 11.11 shows how this can be achieved. Typically an entire output module will be allocated as diagnostic outputs. Each position of the switch will then be used to 'show' one group of output diagnostics at a time.

11.7   In-built diagnostics

Diagnostics are often built onto a system as an afterthought, sometimes by the original designer and sometimes by a third party. Experience over many years in many industry sectors has shown that the diagnostic aids will only work effectively under one of the following conditions: (1) they must be designed into the program from the outset with the same attention to detail as the executive code; or (2) they must be added by the personnel who are responsible for the maintenance of the plant and updated as any omissions or errors are identified.

Experience in industry suggests that the vast majority of maintenance departments are under such pressure to keep plant running that they are rarely in a position to design and update diagnostic code. Some companies have identified this very problem and taken proactive steps to ensure that all programs have built-in diagnostics. Some companies, for instance, require all programs to be written using the EDDI (Error Diagnostic Dynamically Indicated) methodology which forces not only errors to be identified but also a common structure to all programs.

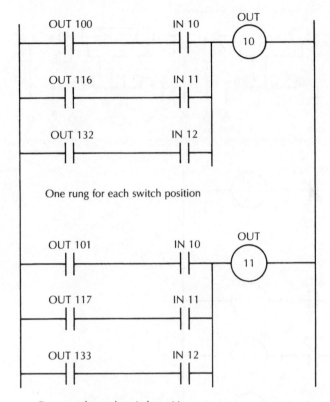

One rung for each switch position

Fig. 11.11 **Use of a single output module to display many diagnostic conditions.**

11.8   Error diagnosis using design documentation

Chapter 7 discussed how the use of motion control diagrams (MCDs) can be used to validate a program design against the requirements. It is also a powerful tool to aid in diagnosing faults on the plant. To illustrate its use we consider the use of the MCD for the following example. Figures 11.12 and 11.13 show the ladder program and MCD for the PLC-controlled circuit A+ B+ C+ C− B− A− as described previously in Chapter 4 (Figure 4.5).

Consider the situation where the system was working correctly and then stopped after operation B+ was achieved. From the diagram it can be seen that the C+ action should have been started by solenoid V3a (OUT 5) being energized because switch 4 was made with the B+ operation. At this particular stage of the sequence all inputs and outputs have a defined status and absence (or more rarely additional, 'rogue') signals can be determined from this diagram. Thus an MCD can present operational information in a more readily understandable form for the maintenance

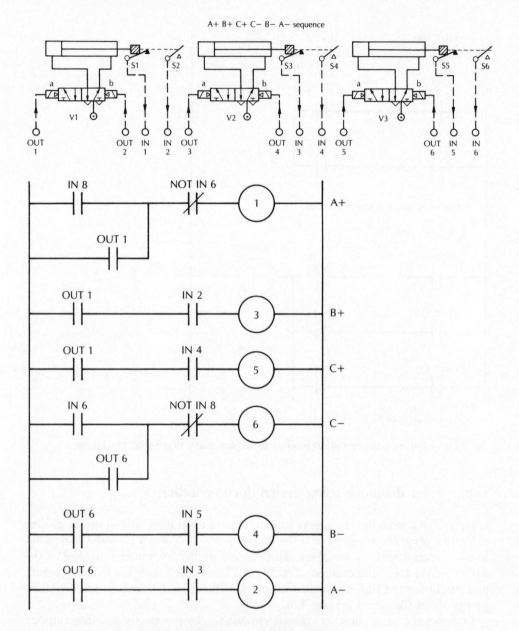

Fig. 11.12 **Example program to illustrate error diagnosis.**

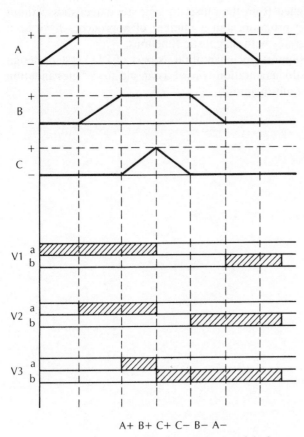

A+ B+ C+ C− B− A−

Fig. 11.13 **MCD used to diagnose operational fault.**

technician to use, since the diagram captures the information contained in the ladder diagram and the details contained in the electrical circuits used with the PLC.

11.9 Summary

The loss of PLC program renders the whole control system useless, therefore software security and support is vital.

The majority of PLCs have input/output light emitting diode displays which are provided to help diagnosis of problems. They are, however, of no use unless the observer has a very good understanding of the program.

It is possible to provide diagnostic help by using diagnostic programs that are designed and programmed into the PLC. A number of techniques are described, each of which is appropriate in different circumstances.

PLC programs can be designed from the outset to generate diagnostics without having an 'add on' diagnostic program, and a number of approaches have been developed to help design programs with diagnostic functions.

It is necessary to generate support documentation as part of PLC system design and installation. The support documentation is used as an aid to design validation and testing and as an aid for fault diagnosis.

CHAPTER TWELVE

Further design exercises

12.1 Exercise 12.1: Paint-spraying system

Figure 12.1 illustrates a paint-spraying system where boxes fed by gravity through a feeder magazine are delivered one at a time onto a moving conveyor belt. A spraying nozzle paints each box as it passes by and a detector D1 counts each box being sprayed. When 20 boxes have been painted and have fallen into the hopper the valve V2 shuts off and cylinder A stops operating.

Eight seconds later the conveyor stops moving and the hopper with its load now moves to the B+ position where it is emptied. Thirty seconds later the hopper returns to the original B− position. End of cycle of operation.

(a) For this sequence of operation design a suitable PLC program.
(b) Suggest a modification to the hardware and the program to prevent the spray system operating if no boxes are present in the feeder magazine.

12.2 Exercise 12.2: Cleaning and paint process

Figure 12.2 shows a cleaning and painting process which is to be controlled by a programmable logic controller. Metal brackets are cleaned in tank 1, then painted in tank 2 using the two air cylinders A and B working in a fully interlocked system. The operation required is as follows:

1 The brackets are hung on the hook above tank 1 (manual operation).
2 The push button START is pressed by the operator.
3 Cylinder A+ action occurs.
4 A bracket is left in the cleaning fluid for a period of 9 seconds to allow cleaning action to occur.
5 Cylinder A− action occurs.
6 Cylinder B− action occurs, placing the bracket to be painted over the top of tank 2.
7 Cylinder A+ action occurs.
8 A delay of 20 seconds occurs to allow painting to be completed by total immersion.
9 Cylinder A− action occurs.

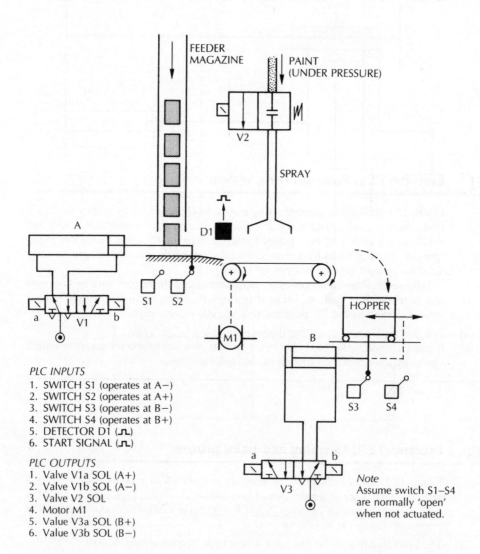

FEEDER
MAGAZINE

PAINT
(UNDER PRESSURE)

V2

SPRAY

A

D1

S1   S2

HOPPER

M1

B

S3   S4

a   b

V1

PLC INPUTS
1. SWITCH S1 (operates at A−)
2. SWITCH S2 (operates at A+)
3. SWITCH S3 (operates at B−)
4. SWITCH S4 (operates at B+)
5. DETECTOR D1 (⊓)
6. START SIGNAL (⊓)

PLC OUTPUTS
1. Valve V1a SOL (A+)
2. Valve V1b SOL (A−)
3. Valve V2 SOL
4. Motor M1
5. Value V3a SOL (B+)
6. Value V3b SOL (B−)

a   b

V3

Note
Assume switch S1–S4
are normally 'open'
when not actuated.

Fig. 12.1 **Paint-spraying system.**

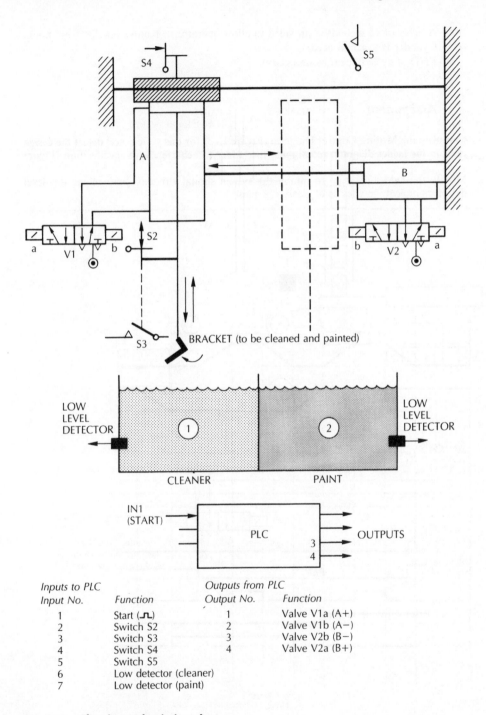

Inputs to PLC

Input No.	Function
1	Start ($\sqcap$)
2	Switch S2
3	Switch S3
4	Switch S4
5	Switch S5
6	Low detector (cleaner)
7	Low detector (paint)

Outputs from PLC

Output No.	Function
1	Valve V1a (A+)
2	Valve V1b (A−)
3	Valve V2b (B−)
4	Valve V2a (B+)

Fig. 12.2 **Cleaning and painting plant.**

10   A delay of 15 seconds is provided to allow operator to remove bracket from hook.
11   Cylinder B+ action occurs.
      END of cycle (repeat as necessary).

Assignment

1   Complete the Motion Control Diagram (Figure 12.3) for this system and **detect the design error in the ladder diagram**, redesigning the ladder to achieve above specification (Figure 12.4).
2   Modify the ladder diagram to inhibit the system should either or both of the low-level detectors indicate low levels of cleaner or paint.

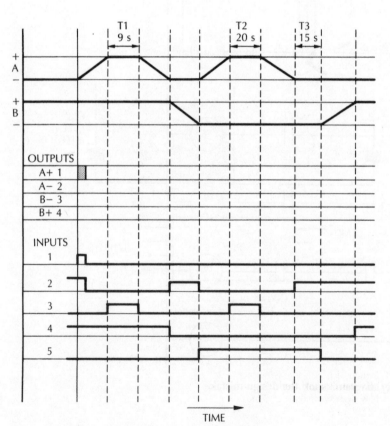

TIME

Fig. 12.3 **Motion control diagram for paint plant.**

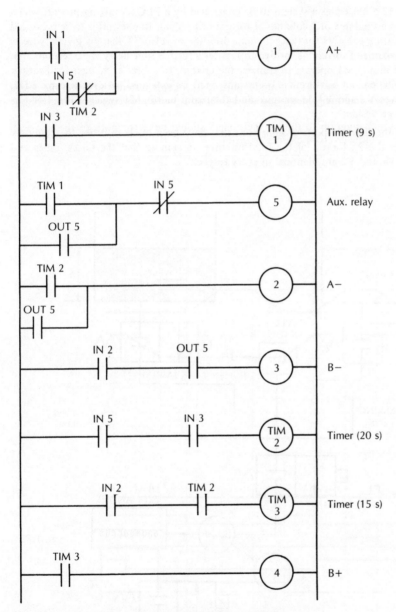

Fig. 12.4 **Ladder diagram: 'spot' the design mistake.**

12.3   Exercise 12.3: Mixing process

Figure 12.5 illustrates a system to be controlled by a PLC. A mixing process which specifies a final mix of 2 volumes of solvent to 1 volume of dyestuff is to be produced by operating valve V1 for twice the time duration of valve V2, thereby giving twice as much volume of solvent as for dyestuff. After a further time delay of 10 seconds, V3, V4 and motor M1 operate to transfer the mixture to a heat exchanger for heating. When 600 cm^3 of mixture has been transferred, as indicated by level detector LD1, the valves V3 and V4 de-energize and the pump motor M1 switches off. For this prototype design:

(a)   discuss the fluid property assumptions that have been made in order to accurately produce the 2 : 1 ratio of the final mixture – assuming that the tanks, pipes and valves V1 and V2 are identical in every respect;

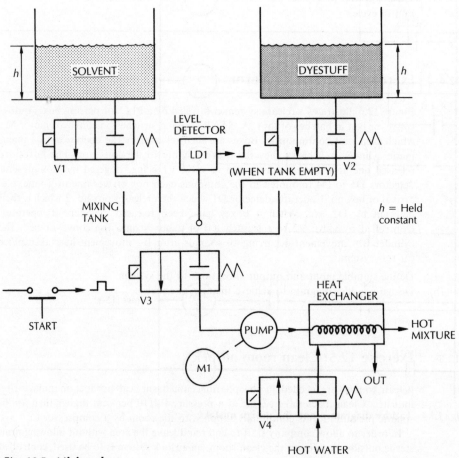

Fig. 12.5   **Mixing plant.**

(b) devise an alternative system to produce the 2 : 1 ratio, not using the time delay method above;

(c) if the flow of dyestuff from the constant head tank is $10\,cm^3/s$, determine the time t needed to produce $600\,cm^3$ of mixture;

(d) from the assumption and figures derived from (a) and (c) above, design a ladder diagram to achieve the process specifications as detailed below:

START
V1 and V2 open
Delay t seconds
V2 closed
Delay $2t$ seconds
V1 closed
10 second delay
Open V3, run M1, open V4 until
LD1 operates (when $600\,cm^3$ has left the mixing tank)
Close V3, stop M1, close V4
End of cycle.

12.4   Exercise 12.4: Box conveyor

Figure 12.6 illustrates a release system controlled by a PLC for placing boxes onto a conveyor belt in batches of four. A 'START' signal switches on conveyor motor M1 which then runs continuously. Boxes are then released by the operation of a pneumatic cylinder A, whence they fall down a chute to the conveyor. The boxes are released one at a time, the command to valve V1 being triggered by 'box present' detectors D1 to D4 (mounted in the chute) as each box arrives one at a time, e.g. arrival of box no. 1 operates detector D1 which then releases box no. 2 which is then detected by D2, etc. When 4 boxes have been released cylinder 'B' operates. Controlled by valve V2 to release batch of 4 boxes onto the conveyor belt, the cylinder B+ movement occurring 10 seconds after B– movement has taken place for this system.

(a) Define suitable input and output allocations to the system.
(b) Design a ladder diagram to achieve the above sequence.

12.5   Exercise 12.5: Clean room airlock

'Clean rooms' are needed in the pharmaceutical–surgical product manufacturing industry. Clean 'bacterial free' air at a pressure P_1 (10 per cent higher than atmospheric pressure, P_A) is continually supplied to the room by a compressor C.

In order to allow company staff to enter and leave the area without allowing non-sterile outside air to enter the clean room, an airlock system is to be used, controlled by a programmable logic controller.

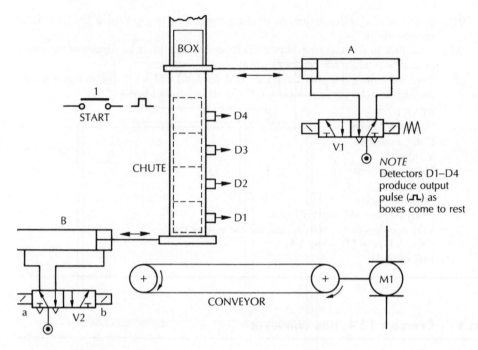

Fig. 12.6 **Box conveyor.**

Operation of system: initial condition, as shown in Figure 12.7; pressure in airlock = P_1.

Sequence

START
1   Person operates SW1
2   Valve V4 vents airlock to atmospheric pressure (P_A)
3   Pressure detector PS detects pressure P_A, V4 'off'
4   Valve V2 energizes, cylinder A operates to open outer door
5   Person enters airlock
6   After a time delay of 10 seconds, V2 'off'
7   Valve V1 energizes, recharging airlock with sterile air
8   Pressure detector senses pressure P_1, V_1 'off'
9   Valve V3 energizes, cylinder B operates to open inner door
10   Person enters room
11   After 5 seconds delay B+ occurs – inner door shuts, V3 'off'

For this system:

(i)   Assign appropriate input/output designation for each item within the system.
(ii)   From the operation specification above, design a PLC ladder diagram that will perform this sequence of events.

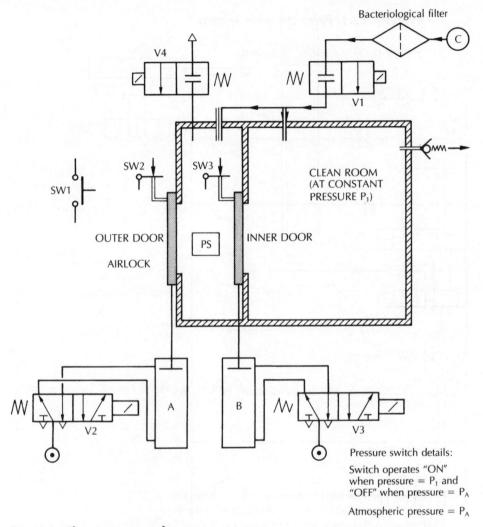

Fig. 12.7 **Clean room example.**

12.6 Sample solutions to exercises

Note. These solutions are *not* the only answers available. Any program that achieves the operational specification is equally correct.

Exercise 12.1: Paint-spraying system

Solution (a) – Ladder diagram

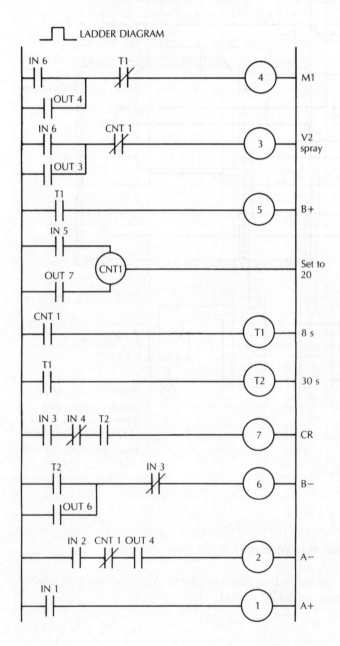

Solution (a) – Motion control diagram

(Note: only 3 actions of A+ A− shown)

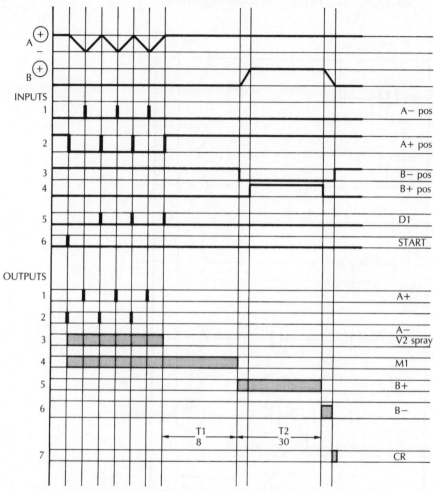

Solution (b)

To prevent the spray operating in the event of no boxes in magazine – add detector D2 situated in magazine (= IN 7) and add to the rung 2 of ladder

Exercise 12.2: Cleaning and paint process

Solution – Correcting the ladder diagram

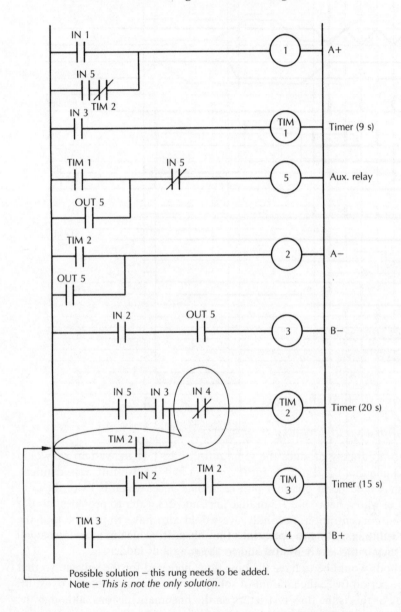

Possible solution – this rung needs to be added.
Note – *This is not the only solution.*

Solution – After the ladder diagram has been corrected

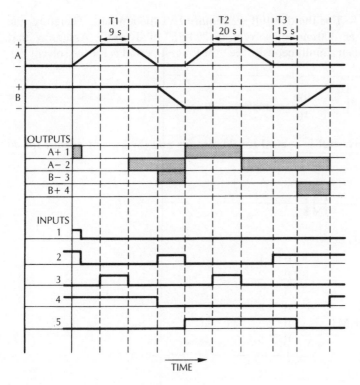

Exercise 12.3: Mixing process

Solutions to parts (a)–(c)

(a) In order to accurately produce the same ratio of 2 : 1 in the system we would have to assume that the *density p*, for the solvent and the dyestuff, was identical. An analogy would be to obtain 1 pint of orange juice, which is twice the concentration of water, hence use 1/3 orange juice and 2/3 water to provide a pint of the required concentration. Secondly, we would also have to assume that the substance will mix. In many operations the substance which is more dense will 'settle out' while the less dense substance 'floats'.

(b) Other methods would be to have the pipe flow increased for the solvent, so that in the same period twice the amount of solvent pours in comparison to dyestuff. An analogy of this is the flow restrictors on the pneumatic pistons, although the input pressures are the same (in the above case gravity). The time it takes to retract or extend can be lengthened or delayed by restricting the flow out. Hence, in the above system, to keep the same period the size of the solvent pipe should be increased.

(c) Flow $= 10\,cm^3/s$. For the dyestuff we require half the amount. Therefore, the mixture would be $400\,cm^3$ of solvent to $200\,cm^3$ of dyestuff. Assuming that valves 1 and 2 open simultaneously, then the longest period is for the solvent, i.e.

$$\frac{400}{10} = time\ (s) = 40\ seconds$$

Therefore, the dyestuff is opened for *20* seconds and the solvent opened for *40* seconds.

Inputs are
LD1   → IN 1
start   → IN 8

Outputs are
V1   — OUT 1
V2   — OUT 2
V3   — OUT 3
V4   — OUT 4
Motor M1   — OUT 5
CR   — OUT 6

Solution to part (d)

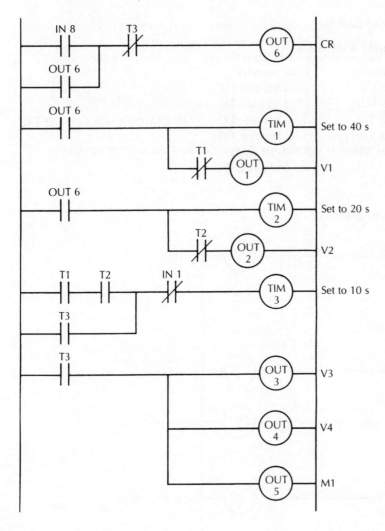

Exercise 12.4: Box conveyor

Solution to part (a)

Input/Output designation

Output number
1 – Valve V1
2 – Valve V2b
3 – Valve V2a
4 – Control relay
5 – Motor M1

Input number
1 – Detector D1
2 – Detector D2
3 – Detector D3
4 – Detector D4
5 – Start signal
6 – Stop signal

N.B. all inputs will give a pulse

Solution to part (b)

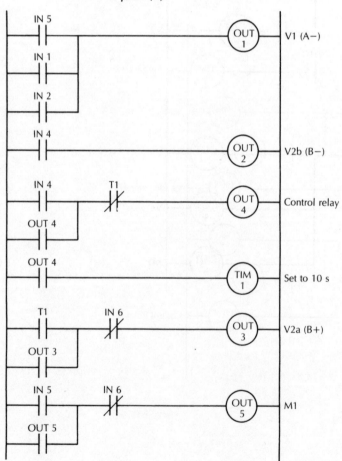

Exercise 12.5: Clean room airlock

Solution to part (i)

Input/Output designation

Output number	*Input number*	
1 – Valve V1	1 – SW1 Start – pulse	
2 – Valve V2	2 – SW2	
3 – Valve V3	3 – SW3	
4 – Valve V4	4 – PS – pressure switch	$\rfloor\ulcorner$ when P_1 present
5 – Control relay		

Solution to part (ii)

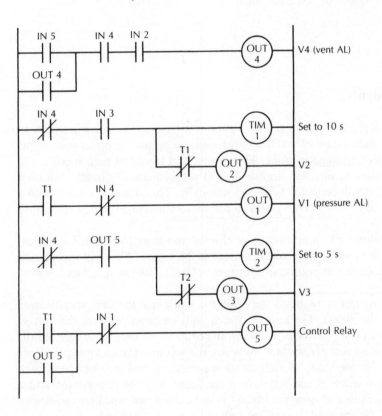

Worked example of PLC size estimation

As an example to illustrate the design and estimation process, we will consider the control system for a piece of test equipment.

A.1   Requirements

A control system is required for a motor torque test (for details, see Figures A.1–A.4). It will run a wide range of 440 V a.c. three-phase motors at up to five torque loads set with an electromagnetic brake. The current and torque at each speed are to be measured and tested against test limits. The test will be manually loaded but once the guard is closed it will continue fully automatically. The motor is loaded onto a fixture which is set by the control system to the correct height to match the motor frame size.

All the test conditions and limits are to be entered and stored with their specification numbers for the entire range of 20 motors to be tested. This will include the fixture height, number of test points, and torques to be set with the speed and current limits for each.

When a change in test is required the operator will enter the new specification number and close the guard. The control system will automatically set the correct fixture height and the test parameter from the stored data. As the test progresses the system will monitor all test parameters, stopping the test and giving an explanatory message if any are outside limits. If all tests are passed the guard is to be opened and the pass lamp illuminated. If the test fails a fail lamp is to be illuminated and a message stating the fault and measured result is to be displayed. In these conditions the guard will stay closed until a specified button or key is operated.

Full calibration facilities must be built into the rig to enable each of the analog channels to be displayed and necessary calibration constants calculated and stored by the PLC.

Upon request, a full set of test results is to be printed at the end of each test.

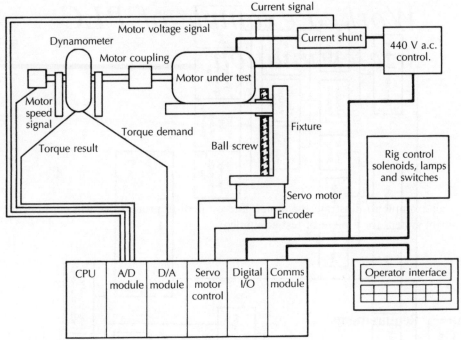

Fig. A.1 **Schematic diagram of motor torque test rig.**

A.2   Memory estimation

This estimate follows the method described in section 9.3.4, the stage numbers referring to that description. A project summary sheet is presented at the end of the estimate.

1 Estimate the effective number of I/O lines. There are 20 digital I/O lines and we need 50 function keys to control calibration, auto, manual, etc.

   Input/output   =   70

2 Scale by a complexity factor. It is estimated each I/O line will be scanned twice in the program.

   Effective I/O COUNT = 2 * 70   =   140

3 Calculate the base size in bytes.

   BASE = 140 * 15   =   2100

4 Calculate the code to handle the complex I/O (4 A/D, 1 D/A, 1 encoder).

   ANALOG   =   1000

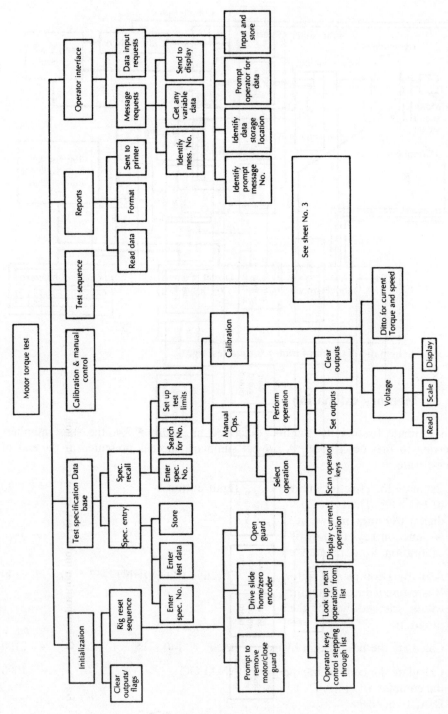

Fig. A.2   **Decomposition for motor test.**

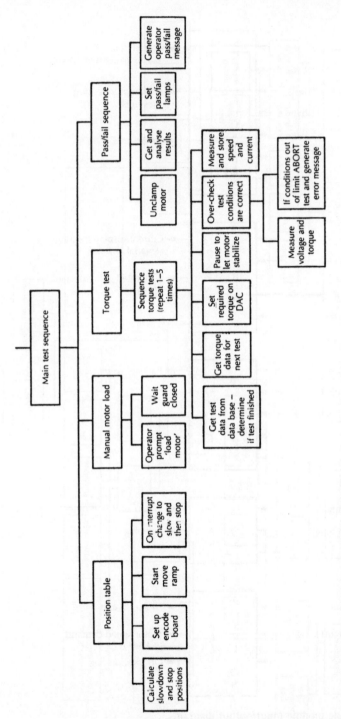

Fig. A.3 **Decomposition motor test: main test section.**

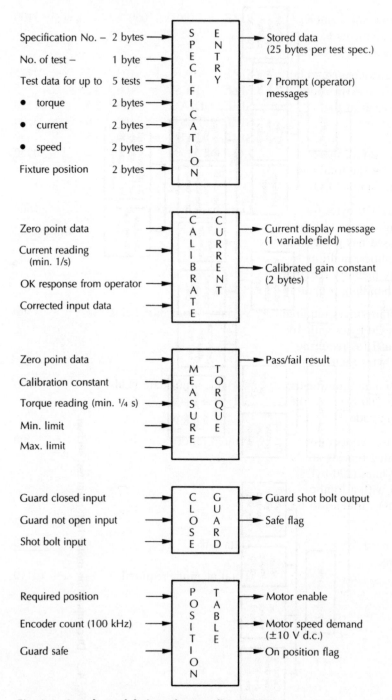

Fig. A.4 **Sample module input/output diagrams.**

5 Add the simple and complex memory requirements to determine the memory SIZE.

$$\text{SIZE} = 2100 + 1000 \qquad = 3100$$

6 Scale the size calculated so far to determine the extra code required for each of the following features:

$$
\begin{aligned}
\text{SIZE} \times \text{MAN} &= 3100 * 0.25 &&= 775 \\
\text{SIZE} \times \text{REST} &= 3100 * 0.25 &&= 775 \\
\text{SIZE} \times \text{DIAG} &= 3100 * 0.5 &&= 1550 \\
\text{SIZE} \times \text{TEST} &= 3100 * 1.0 &&= 3100 \\
\text{SIZE} \times \text{CAL} &= 3100 * 0.5 &&= 1550 \\
\text{I/O TOTAL} & &&= 10850
\end{aligned}
$$

We can now add all these segments to get the total code to handle the 'simple' I/O.

7 The rig needs 830 bytes for data and flag storage. As the PLC to be used has good data-handling facilities, multiply by 2 to calculate memory required for data + handling programs.

Data base memory $= 1660$

8 There are 60 messages required, each needing 20 bytes with 10 variable data fields requiring an extra 30 bytes each.

Variable input data fields $= 1500$

9 There are 10 data input fields, which equals 200 bytes + the data handling code

Variable input data fields $= 600$

10 In addition to the operator display, printed reports are required. 10 lines (200 bytes) are required with a total of 10 (300 bytes) variables to be printed.

Printed report $= 500$

11 There is no need for a computer link

Data link $= 0$

Total memory required $= 15110$

Project summary sheet

Job name __*Motor test rig*__   Engineer __*W. Jeffcoat*__   Date __*30/6/92*__

Digital inputs.   D.C. voltage __24__   No. __13__   voltage _____ No. _____
Digital inputs.   A.C. voltage _____ No. _____ voltage _____ No. _____
Digital outputs.   D.C. voltage __24__   No. __7__   voltage _____ No. _____
Digital outputs.   A.C. voltage _____ No. _____ voltage _____ No. _____
Analog inputs.   No. __4__ max. voltage __10__ Accuracy (bits) __12__ Uni-or bipolar __U__
Analog outputs.   No. __1__ max. voltage __10__ Accuracy (bits) __12__ Uni-or bipolar __U__
Counter inputs.   No. _____ voltage _____ Speed (counts/s) _____
Encoder inputs   No. __1__ voltage   __5__ Speed (counts/s) __100 kHz__
Servo control   No. __1__ encoder rate Voltage out __10__ positioning accuracy __±1 mm__
3-term control   No. _____ voltage _____ Voltage out _____ min. update rate _____ (s)
Other 'special' modules _____
'BASIC' module   Mem. size _____ comms port type _____
Communications module   Mem. size __4K__ comms port type __RS232__
Advanced program requirements _____ Data handling __*Data base*__
                                         Maths functions __*Analog scaling*__
                                         Floating point _____
Estimated program size _____
Targeted program scan rate _____ __*50 ms*__ _____
Preferred suppliers __*Siemens*__   *A.B.*
Any special methods required (Grafset; EDDI; etc.) _____

Index